TECHNOLOGY'S DILEMMA

AF776318

Jonathan Harwood

TECHNOLOGY'S DILEMMA

Agricultural Colleges between Science and Practice in Germany, 1860–1934

PETER LANG
Oxford • Bern • Berlin • Bruxelles • Frankfurt am Main • New York • Wien

Bibliographic information published by Die Deutsche Bibliothek
Die Deutsche Bibliothek lists this publication in the Deutsche Nationalbibliografie; detailed bibliographic data is available on the Internet at ‹http://dnb.ddb.de›.

British Library and Library of Congress Cataloguing-in-Publication Data:
A catalogue record for this book is available from The British Library, Great Britain, and from The Library of Congress, USA

Cover design: Thomas Jaberg, Peter Lang AG

ISBN 3-03910-299-0
US-ISBN 0-8204-6987-4

© Peter Lang AG, European Academic Publishers, Bern 2005
Hochfeldstrasse 32, Postfach 746, CH-3000 Bern 9, Switzerland
info@peterlang.com, www.peterlang.com, www.peterlang.net

All rights reserved.
All parts of this publication are protected by copyright.
Any utilisation outside the strict limits of the copyright law, without the permission of the publisher, is forbidden and liable to prosecution.
This applies in particular to reproductions, translations, microfilming, and storage and processing in electronic retrieval systems.

for my mother, Virginia Oakes Tyler, both sides
of whose family still farm in western New York

Contents

Abbreviations

BayHStA	Bayerisches Hauptstaatsarchiv, Munich
GstaA	Geheimes Staatsarchiv Preussischer Kulturbesitz, Berlin
RAC	Rockefeller Archive Center, Tarrytown, New York
R21	(Reichsministerium fuer Wissenschaft, Erziehung und Volksbildung), II. Anhang (Erfassung der Hoch schullehrer, 1934–44), Bundesarchiv Koblenz
TUM-Archiv	Main Archive of the Technische Universitaet Muenchen
UA-Giessen	Archive of the Universitaet Giessen
UA-Halle	Archive of the Universitaet Halle
UA-Hohen	Archive of the Universitaet Hohenheim
UA-Weih	Archive of the Technische Universitaet Muenchen Verwaltungsstelle Weihenstephan, Freising
Wuertt HstaA	Wuerttembergisches Hauptstaatsarchiv, Stuttgart

Tables and Figures

Preface

'Technology's dilemma' is rooted in its very nature. Like most of us, technology walks on two legs. On the one hand, since the 19th century technologists – above all those who work in academe – are expected to keep an eye on the basic sciences for concepts, theories, methods and so forth which may be of use in fashioning improved technologies. On the other, however, a technology must be suited to the needs of its destined users: it should be effective, easy to use and affordable. But for reasons that will become clear in this book, it has not been easy for academic technologists to walk evenly on both of these legs. The pressures to pay more attention either to the sciences or to user-groups are considerable. It is very common, therefore – in agriculture as in other areas of technology – to find some colleges whose teaching and research are more strongly inclined toward science while others are more practically oriented. The central problem in this book is to understand how these differences of orientation have come about and why 'practical' institutions have tended to drift closer to the scientific pole over time.

When I began writing, my intended audience included historians of agriculture and the agricultural sciences as well as those in science and technology studies who are interested in the social construction of technical knowledge. But I also wanted to reach a wider range of readers. For one thing, the 'dilemma' in question arises in a number of disciplinary areas including engineering, medicine and management sciences since all of these areas are obliged to operate in that uncomfortable middle-ground between 'science' and 'practice'. But it is also clear that technology's dilemma bears upon educational policy issues. For in recent decades educators in several of these areas have expressed concern that technical and professional education has drifted too far toward 'science', such that graduates lack basic practical skills, and research is out of touch with the needs of industry, farmers or medical practitioners. This phenomenon of 'academic drift' is a central theme in the book, and in the concluding chapter I draw out the implications of my analysis for the reform of technical education.

By the time I had finished writing, however, I realised that *Technology's Dilemma* was also dealing with processes of interest to yet another audience. As is well known, large sections of the public in several Western countries are currently concerned about food safety – not least in Britain, prompted by recent crises over BSE and the foot-and-mouth epidemic – as well as farming's impact upon the environment. As I write, European governments are having to take decisions whether to allow the commercial planting of genetically modified crops, and the World Trade Organisation is seriously divided over whether the high level of European and American farm-subsidies is defensible. In one way or another, all of these issues are a product of high-tech or 'industrial' agriculture in the West. And this is the approach to agriculture which was taught in the 'science-oriented' colleges central to my story. As will become apparent in the following chapters, however, science-based agriculture – which has been focussed primarily upon productivity and tailored to the needs of large landowners – did not go unchallenged. At several German agricultural colleges the staff took a different line, developing a more 'practically' oriented curriculum and addressing their research to the needs of small-holders (who constituted the great majority of the farming community). Nevertheless, after the First World War practice-oriented agricultural education was gradually eroded as science-based colleges became the norm. *Technology's Dilemma* is thus in part a study of the early promotion of industrial agriculture as well as of 'the road not taken'.

The book's origins go back to the mid-1980s when I and two colleagues, R.D. Whitley and John Pickstone, obtained funding from the (British) Economic and Social Research Council for a research fellow (Colin Divall) to study the history of British engineering education 'between science and industry'. The results of this work were interesting, but with no background in engineering, I decided to explore these tensions between 'science' and 'practice' further in an area in which I was more at home. Though at the time I knew nothing about agriculture, I was finishing a book on the history of genetics,[1] and it seemed promising to turn to an area of the agricultural sciences in which genetics might have been central. Thus in the late '80s I and an agricultural economist, Colin Thirtle, approached the ESRC for another grant to look at the history of public-sector plant-breeding in Britain. It was this project which triggered my interest in agriculture more generally, and I learned a great deal from discussions with Paolo Palladino, the research fellow whom we appointed. Stimulated by his work, I began to look at the history of plant-breeding in Germany but soon realised that the issues of 'science' and 'practice' went far beyond plant-breeding; it would really be necessary to study agricultural education as a whole. Some of my work on plant-breeding has found a place within this wider project (see Chapter Four), but the rest will have to wait until another book.

Since beginning the research for this project in 1989, I have had help from a wide variety of sources. Archival research was possible, thanks to several small travel grants from the Royal Society as well as a stipend from the Deutscher Akademischer Auslandsdienst. In 1994–95 I was able to make headway on the project while a fellow at the Dibner Institute for the History of Science and Technology at MIT, supported by a stipend from the Institute as well as a Senior Research Fellowship from the ESRC (award no. H52427502694). Since then additional archival research has been conducted while I was a guest of the History and Philosophy of Science Program at the University of California, Davis (with thanks to Michael Dietrich and James Griesemer) as well as of the Forschungszentrum am Deutschen Museum in Munich (with thanks to its director, Helmuth Trischler). Much of the writing was completed during several stays at the Max-Planck-Institut fuer Wissenschaftsgeschichte in Berlin, and I am indebted to Hans-Joerg Rheinberger for making this possible. Lastly, I am grateful to Michael Worboys and my home department (the Centre for the History of Science, Technology and Medicine, University of Manchester) for covering the costs of publication.

Work of this kind is heavily dependent on knowledgeable (and cooperative) archivists and librarians, several of whom went out of

1 Harwood, *Styles of Scientific Thought.*

their way to provide useful advice and other assistance. I would especially like to thank Dr. Ulrich Fellmeth (Universitaetsarchiv Hohenheim), Dr. Martin and Frau Sante (Universitaetsbibliothek Hohenheim), Frau Margot Fuchs (Historisches Archiv der Technischen Universitaet Muenchen), Herr Rudolf Lackermeier and his colleagues (Verwaltungsstelle Weihenstephan der T.U. Muenchen), and Peter Filby (Scientific Periodicals Library, Cambridge University).

Colleagues have been helpful in a variety of ways. Andreas Heisler helped me to find my way around piles of dusty papers in the basement of Munich's Technische Universitaet in the days before its Archive was formally constituted. Stewart Richards provided a copy of his stimulating book manuscript on the history of agricultural education in Britain, and I benefitted, too, from Keith Tribe's knowledge of German business education. For help of various kinds I also thank Mark Finlay, Deborah Fitzgerald, Peter Gatrell, Jean Gayon, Richard Hankins, Roland Harwood, Barbara Kimmelman, Bob Olby, Paolo Palladino, Diane Paul, Nils Roll-Hansen, and Thomas Wieland. Rob Kohler's (characteristically trenchant!) criticisms of the entire manuscript have made it a much better book. For valuable advice and encouragement over the last ten years I am also grateful to Wolfgang Boehm (Goettingen) and Volker Klemm (Berlin). Without their ground-breaking work on the history of German agricultural sciences, writing this book would have been far harder.

Parts of the argument were presented at colloquia in 1995 at the Dibner Institute for History of Science and Technology at M.I.T., the Department of Science and Technology Studies at Cornell University, and the Department of History of Science at Harvard University. In the autumn of 1999 I gave papers based on this project at the Department of History and Philosophy of Science at Cambridge University, the Deutsches Museum in Munich, and the Department of History and Sociology of Science at the University of Pennsylvania. The feedback from colleagues at those institutions – and especially those at Penn with their strengths across the entire field of technology, medicine and science – was much appreciated.

Manchester
August 2004

Chapter 1: Technology's Dilemma

Technical education is evidently in trouble. Over the last generation interested parties have complained that technical and professional education, in several fields and in several countries, is out of touch with 'the real world'. Industrialists in the United States, Britain and France, for example, have objected that the curriculum and research in engineering colleges are excessively theoretical.[1] In medicine several reports by American medical organisations during the 1980s called for less time in the curriculum to be devoted to specialist knowledge and laboratory sciences and more to be spent on teaching basic practical skills. In Britain similar concerns from the 1990s have lead to the reform of medical education at several institutions.[2] And since the 1970s American agricultural colleges have been criticised for turning their backs upon the family farm and the rural communities which they were created to serve.[3] In each of these cases critics are concerned that technical education has drifted away from its originally practical orientation, becoming increasingly irrelevant to actual needs.

How might one set about solving this problem? If such cases of 'academic drift' are to be reversed (and avoided in future), we obviously need to understand how institutions of technical education work. That is, what processes affect the evolution of the curriculum in such institutions? How are research priorities set? Who has a say in the kinds of staff who are appointed? What factors tend to encourage academic drift (and conversely, how is it that a few institutions nonetheless manage not to drift)?

The natural place to turn for answers would be to the history of engineering, medicine and agriculture since these literatures contain numerous studies of the relevant educational institutions. But few

1 Seeley, 'Research, engineering and science', 383; Ferguson, *Engineering and Mind's Eye*, 159–168; Williams, 'All that is solid'; Divall, 'Fundamental science versus design'; Grelon, '*Die deutschen Ingenieure*', 383.

2 Rothstein, *American Medical Schools*, 300–302, 312–313, 326–327; Mahadevan, *Creating Tomorrow's Doctors*.

3 Hightower, *Hard Tomatoes*; Hadwiger, *Politics of Agricultural Research*.

answers will be found. For although the phenomenon of academic drift is well-recognised, there have been very few attempts to account for it. Indeed, the problem is more fundamental; these literatures have little to say about the social and intellectual dynamics of such institutions. And without models for understanding how research or the curriculum evolves, we cannot explain why most of these institutions drift while a few do not. My aim in this book is to develop just such a model.

Practical Sciences in No-Man's-Land

At the outset we need to be clear about the scope and the nature of the problem. Just which fields are affected by academic drift and why? I have just referred to technological fields like engineering and agriculture as well as to professional ones like medicine, but even this relatively broad domain is not broad enough. As I will argue in the concluding chapter, for example, the cluster of disciplines known as the 'management sciences' also displays many of the same phenomena. Since historians are not used to discussing a domain which embraces so many diverse disciplinary groups, we need a term to designate all of these fields and others like them. I will, therefore, refer to this cluster henceforth as the 'practical sciences'.

What all of these fields share is that they are expected to draw upon systematic bodies of fundamental knowledge – whether in the natural or social sciences or even in the humanities – in order to solve practical problems (ie, those arising in the economy or in other areas of our daily lives).[4] Thus the signal feature of the practical sciences is that they have dual allegiances: they must, as it were, look in two

4 Unfortunately this cluster of disciplines has often been labelled the 'applied sciences', implying that their practice consists of applying theory and concepts from the 'pure' sciences to practical problems. But this clearly won't do; as many studies in the history of technology over the last few decades have demonstrated, this is an inadequate conception of what problem-solving in these fields actually consists of (eg, Staudenmaier, *Technology's Storytellers*).

directions at once or juggle two quite different priorities. They must at one and the same time ask 'what can we learn from the basic disciplines?' and 'how can we devise solutions to practical problems?' This is not always a comfortable position to be in, and the problem is particularly acute for the academics who work in these fields.[5] For although trained (at least in part) in basic disciplines and often employed alongside scholars in those disciplines, practical scientists have rarely been accepted as fully-fledged members of the scholarly community.[6] Moreover while at the same time practical scientists are expected to serve a wider economic constituency, this constituency has often mistrusted them, reckoning them to be in the ivory tower and thus out of touch. Thus the practical scientist is stuck betwixt and between, marginal to two relatively powerful groups and inviting damnation by the one, should he court the other.[7] This is the 'dilemma' to which my title refers.

How do people cope with conflicting pressures of this kind? Common sense tells us that in the search for acceptance, they often throw in their lot with one side or the other, though occasionally a few will try to straddle the divide. This is, in fact, just what one finds with the practical sciences. At one extreme are those whose approach to problem-solving was what I will call *science-oriented*. These individuals sometimes took their research questions from the basic sciences and attempted to solve them through application to data or problems from the practical realm. Alternatively, they took their research problems from the practical realm but were convinced that their solution lay merely in the correct application of scientific theory.

5 Early 20th century American farm management experts provide a nice example. Though academically trained, many came from farm backgrounds so that when addressing agricultural problems, 'it was a bit of a balancing act to maintain their scholarly credentials while drawing on their experience in the field, because their work and their identity were split between the two sites [...]' (Fitzgerald, *Every Farm a Factory*, 43–44).

6 Medical academics are probably an exception in this respect.

7 I use the term 'he' advisedly; German agricultural science into the early 20th century was a virtually all-male occupation. As late as 1925, for example, only one percent of agriculture students were women (Titze, *Datenhandbuch*, Vol. 1, pt. 1, table 69).

Or they were enamoured with the methodological apparatus of the sciences, making a point in their research or teaching to deploy the techniques, concepts, laws, data, and instruments from one or other basic discipline. In so doing, these academics were effectively seeking recognition primarily from the basic science community. An example of this category might be the Norwegian, Vilhelm Bjerknes, who, during the early part of his career in meteorology, drew upon his original training in theoretical physics in order to develop a precise and quantitative mechanics of the atmosphere from which he hoped it would be possible to predict future atmospheric states.[8]

At the other end of the spectrum were those academics whose approach was *practice-oriented.* This group generally took their research questions from the practical domain, and in attempting to solve them, they often drew *in part* upon theories and methods from the basic disciplines, though without assuming that these alone would be sufficient for a solution. Instead these academics were attentive to what was both economically feasible and practically realistic under the particular circumstances in which their 'clients' in the practical domain were working. (One example of this type is the British colonial entomologist who remarked that on encountering an agricultural problem, the first thing one needed to study was not the insects but the *crops* in question, along with local cultivation practices and the resources of the farmers concerned.[9] Another example is the design-engineer whose inventions must take into account not only scientific and technical possibilities but also the practical circumstances in which the artifact will be manufactured and used as well as economic – and sometimes aesthetic – constraints.) On other occasions, however, solutions were derived merely from empirical experience or from modifying existing technical practices with no recourse to basic

8 Interestingly, as Bjerknes became involved in weather forecasting, he gradually abandoned his earlier theoretical aims, and his approach to meteorology became less deductive and more empirical and qualitative in order that he could better meet the needs of farmers, fishermen and the military. In my terminology, therefore, he moved from a science-driven approach toward a practice-driven one. (Although I have taken this example from Kranakis, 'Hybrid careers', her account is based on Friedman, *Appropriating the Weather*).

9 Jacob, 'Reflections of an applied biologist'.

knowledge. Common to both of these approaches was the subordinate role assigned to basic disciplines in the problem-solving process. Thus practice-oriented scientists were evidently seeking recognition primarily from a wider non-academic constituency.

Occasionally, one finds academics who attempted to tread the tightrope, interested in problems within basic disciplines but also attending to the needs of an economic clientele so as to come up with knowledge or techniques which not only provided practical solutions but also secured recognition from the basic scholarly community. One area in which this can occur is where phenomena are equally interesting to both basic scientists and those in the world of practice (an area sometimes called 'strategic research'). In both solid-state physics and molecular genetics, for example, a fundamental understanding of certain processes has also been of commercial interest. The physicist-chemist-engineer, Irving Langmuir, provides a nice illustration of this since his work at the General Electric Company's research laboratory before the First World War sought not only to improve lamp design but simultaneously to develop a theoretical understanding of how lamps functioned (and in the process won him a Nobel Prize).[10]

But we need to be clear about the meanings of 'science' and 'practice'. When I describe individuals' approach to problem-solving as 'scientific', I do not mean that they were necessarily uninterested in practical problems. On the contrary, many were, but the point is that they regarded science as by far the most important element in the solution to such problems. Conversely, when I describe individuals' orientation as 'practical', I do not mean that they were either ignorant of – or made no use of – theories, laws, concepts etc. from basic disciplines. Instead they regarded these disciplines as just one resource among many for the solution of practical problems. In the language familiar to historians of technology we might say that the 'scientific' approach embodied the notion of technology as applied science while the 'practical' one regarded design as the essence of technical innovation. From another angle, the 'scientific' and 'practical' approaches also embodied differing conceptions of the proper relationship between academe and user-groups in the wider economy. For an indi-

10 Reich, *Making of American Industrial Research*, chp. 5.

vidual whose orientation was 'scientific', the nature of problems and acceptable solutions was defined primarily by academics, users playing a relatively passive role. For the 'practically' oriented scientist, in contrast, users played a far more active role, defining both problems to be solved and workable solutions.

Consider some concrete illustrations. If we look at the language used by mid-19th century engineers, for example, we see that they were very much aware of the tensions between the world of science and the economy. In Germany, for example, both engineers and industrialists referred to the tendency within academe to focus upon 'Theorie' rather than 'practice' (Praxis). In Britain, too, 'theory' and 'practice' were central to W.J.M. Rankine's discussions of engineering as well as to the 'electrical debate' ca. 1890. As one British engineer put it, '[the engineer] is a mediator between the Philosopher and the working Mechanic; and like an interpreter between two foreigners must understand the language of both'.[11] Sometimes this tension was also found within the academic world. In the German debate among engineering professors during the 1890s over the proper role of mathematics in the engineering curriculum (viz, whether or not it was to be taught in a manner which was tailored to engineering needs) the two main factions were referred to as 'Theoretiker' and 'Praktiker'.[12] In the United States, similarly, a defender of the new Georgia Institute of Technology's practically-oriented curriculum declared:

> The head is in Athens [ie, Athens, Georgia, site of the University of Georgia which had a stronger liberal–arts tradition]; the hands are here. We have here thought versus work, practice against theory; the shop against the study; the hammer against the book [....].[13]

11 Cited in Marsden, 'Engineering science', 322. On the electrical debate see Hunt, 'Practice vs. theory'.

12 Hensel, 'Auseinandersetzungen', passim.

13 Nathaniel Harris in 1888, cited by Brittain and McMath, 'Engineers and the New South', 136. Similarly, an academic at a German engineering college who defended a central role for advanced mathematics in the curriculum described himself as a 'theoretician' (Hashagen, 'Der Mathematiker', 280). 'Theory' and 'practice' were common usage among U.S. electrical engineers by the 1890s (McMahon, *Making of a Profession*, 67). On Germany see Gispen, *New Profes-*

In Germany the vocabulary of 'theory' versus 'practice' was also used to describe medical curricula from at least the mid-18th century and was not uncommon in early 20th century management science.[14]

Turning to the question of status, there is no doubt that engineers in various countries since the 19th century have been frustrated by their failure to secure adequate recognition, either from the basic science community or from the world of industry. As Edwin Layton remarked, 'American engineers have been concerned, almost to the point of obsession, with the low prestige of engineering; an identification as "scientist" was expected to raise status.'[15] At the same time, however, late 19th century academic engineers in France, Britain and Germany ran up against scepticism from the industrial community

sion, Old Order, 124; Zweckbronner, 'Ingenieurwissenschaften', 209; Koenig, *Kuenstler und Strichezieher*, 56. On Rankine see Marsden, 'Engineering science'.

14 Broman, 'Rethinking professionalisation'. The title of one of its journals reflected the new management science's attempt to straddle this divide: *Zeitschrift fuer Handelswissenschaft und Handelspraxis* (journal for the science and practice of commerce) (Locke, *End of Practical Man*, 223).

15 Layton, 'American ideologies'. On the status concerns of American engineers, see Seeley, 'SHOT'; Reingold and Molella, 'Introduction': Ferguson, *Engineering and Mind's Eye*, chp. 6. On German engineers' preoccupation with status, see Manegold, *Universitaet, Technische Hochschule*; Hortleder, *Gesellschaftsbild des Ingenieurs*; Zweckbronner, 'Ingenieurwissenschaften'; Jarausch, *Unfree Professions*. Mid-19th century French engineers resented being looked down upon by scientists (Weiss, *Making of Technological Man*, 158–159), and some observers have noted a similar subordination of utility to purity in French engineering education into the 1980s (Day, *Education for Industrial World*, 118ff).
The same status-question has concerned a wide range of practical scientists since the 19th century. On veterinary medicine and forestry, see Moraw, 'Organisation u. Lehrkoerper', 46; Moraw, 'Humboldt in Giessen'; and Baumgarten, *Vom Gelehrten zum Wissenschaftler*, 89. Since chemistry was early perceived to be a practical science, chemists faced the same condescension; one 19th century historian dismissed his colleagues in chemistry as 'apothecaries and manure-drivers' (Radkau, *Technik in Deutschland*, 159; see also Borscheid, *Naturwissenschaft, Staat und Industrie*). In early 20th century Germany some psychologists were loathe to declare their institutional independence from philosophy for fear that their new discipline would then be regarded as a lowly vocational subject (Ash, *Gestalt Psychology*, 47).

who thought their knowledge and curricula were too impractical.[16] Occasionally engineers recognised how they were caught between two worlds and complained of their predicament:

> Standing right in the middle between the scholarly professions, artists, merchants and businessmen [... the engineer] occupies a position in the world whose precariousness and uncertainty are disproportionate to its significance for the structure of society.[17]

And at Oxford in the early 20th century Robert Southwell, noting that engineers were defensive, ascribed it to the fact that their status was uncertain because they seemed to be 'neither plumbers nor scientists'.[18]

A number of historians have recognised this ambivalent social location. Many years ago Herbert Shepherd described the engineer as a 'marginal man', torn between 'the scientific culture' and 'the business culture'. More recently Eda Kranakis observes that 19th century French engineers 'were pulled in two directions', working day to day on technical problems but also seeking recognition from the Academie des Sciences for their theoretical achievements. As Manfred Spaeth puts it, engineers have occupied a 'hybrid locus' (Zwitterstellung) between craft-work and science. John Servos notes that engineers at

16 Grelon, 'Training and career structures', 52, 56; Hunt, 'Practice versus theory'; Mauel, 'Der VDI', 106; Dienel, 'Der Muenchener Weg'. Nor were German businessmen much impressed by business schools before the Second World War (Franz, *Zwischen Markt und Profession*).

17 A mid-19th century railway engineer, cited in Gispen, *New Profession*, 45. In 19th century Germany the perception that engineering was neither chalk nor cheese was reflected in a term which critics sometimes used when objecting that engineering colleges were trying to transmit *both* scientific and practical instruction but doing neither well. Such institutions were dismissed as a *Halbheit*, literally 'halfness', meaning an incomplete entity or an incapacity to carry out a task thoroughly (Schnabel, *Deutsche Geschichte*, 323).

18 Morrell, *Science at Oxford*, 99. Obviously to note the ubiquity of 19th century engineers' status-unease does not mean that this unease was equally pronounced in every country at that time, nor that it persists unchanged today. That said, conversations with German colleagues suggest that a status-gap still exists today in some places between the technical universities (former engineering colleges) and the traditional universities.

the University of Michigan had to keep businessmen satisfied with their graduates while at the same time maintaining their status within the University. And as Edwin Layton concluded in his history of the profession, the engineer's problem was his marginality; though expected to be both scientist and businessman, in fact he was neither.[19] Although their insight is a valuable one, none of these historians seems to have developed it further.

If we are to understand why academic drift occurs, however, it is crucial to find out why practical scientists have so often preferred to deal with their status-dilemma by developing science-oriented approaches to problem-solving rather than practice-oriented ones. Occasionally it is suggested that the explanation might be a psychological one. When some engineers devote their energies to designing more efficient or 'technically sweet' devices while others aim for devices which can be manufactured more cheaply or are more convenient for the user, some would say that this is just a reflection of personality differences and has nothing to do with the fact that the engineering *occupation* itself is betwixt and between. Such individuals undoubtedly exist.[20] But if differences of approach were purely an individual matter, one would expect all educational institutions for the practical sciences to be inhabited by individuals displaying a wide range of problem-solving strategies, some of them 'scientific' and others 'practical'. And this is not what one finds. Historians of engineering, medicine and agriculture have noted the existence in several countries of substantial differences from one college to the next in the prevailing emphasis

19 Shepherd, 'the engineer and his culture', 211; Kranakis, *Constructing a Bridge*, 263; Spath, 'Ingenieur als Buerger', 84; Servos, 'Engineers, businessmen'; Layton, *Revolt of the Engineers*, 2.

20 On the eve of the First World War, for example, the electrical engineer, William McClellan drew a distinction among three types of engineer: the practical/ mechanical, the theoretical/scientific and the designer (whom he took to be the 'real engineer'; Layton, 'American ideologies', 696). Eda Kranakis discusses individual engineers whose work represented a 'hybrid' of science and technology (Kranakis, 'Hybrid careers'). Hybrid research strategies of course were not restricted to engineering; characteristic of all of Louis Pasteur's work from the 1860s, according to Andrew Mendelsohn, was his aim to unite 'theoretical and utilitarian concerns, bringing together new science and traditional knowledge' (Mendelsohn, *Cultures of Bacteriology*, 24).

upon 'science' versus 'practice', in the curriculum as well as in research priorities. Few of them have gone on to ask where such institutional differences come from, but that is the central question which I will address in the chapters which follow. And in order to do so, I will focus on an area of the practical sciences about which relatively little is known: agriculture.

Whatever happened to the history of agriculture?

Agriculture is very much on the political agenda. Throughout the industrialised world in recent years concerns have been expressed about food-safety, environmental pollution from agrochemicals, and the uncertainties surrounding the introduction of genetically-modified crops. But a moment's reflection shows that agriculture's importance for the West is more fundamental and longstanding. As recently as the mid-19th century, for example, the majority of European and North American populations lived off the land. And as industrialisation got underway during the latter half of the century, increased agricultural productivity – based on techniques derived from the agricultural sciences – was essential: factory workers had to eat. In 'third world' countries as well agriculture remains the foundation of their economies. There, too, agricultural sciences have played a major (and controversial) role since the 1940s, most notably in the form of the 'Green Revolution' which sought to solve the world hunger problem through the introduction of new high-yielding plant varieties. Nevertheless periodic food shortages in that part of the world remain a serious problem and have prompted extended debate over various ways in which the advantages of high-tech agricultural methods can be introduced more effectively. All of this means that how farmers and agricultural officials have been educated, the kinds of research undertaken by agricultural colleges, and the links between these institutions and a variety of user-groups are absolutely central to the global economy, then as now.

Given the importance of these issues, however, it is remarkable – to put it mildly – how little attention historians have paid to this subject. Among general historians agriculture ranks among the least popular specialties, and in the history of technology a tiny proportion of the literature has been devoted to agricultural sciences, practices or education. Several academic publishers who during the 1980s ran book-series on agricultural topics no longer have them, and others are concerned that books on the subject will not sell.

That is not to say that none of these issues has been addressed by historians. There is indeed a literature on the history of the agricultural sciences and education, some of it very good, but it is remarkably small. To my knowledge, for example, there are but two book-length historical overviews of the agricultural sciences as a whole (both of them, as it happens, with a German focus).[21] As far as disciplinary history is concerned, the situation is better, but the great majority of this work consists of articles dealing with the basic sciences relevant to agriculture – eg, genetics, entomology, bacteriology – rather than with specifically agricultural disciplines (eg, agronomy, breeding, soil science).[22] Turning to institutional history, there is relatively little on

21 Reichrath's *Entstehung, Entwicklung* is a relatively short dissertation which attempts to compare the emergence of the agricultural sciences in Germany and France over two centuries while relying heavily on secondary sources. Klemm's *Agrarwissenschaften in Deutschland* is a much more substantial work.

22 On genetics in agricultural contexts see Kimmelman, *Progressive Era Discipline*. The agricultural dimensions of microbiology are one of several foci in Vernon, 'Pus, sewage' and Mendelsohn, *Cultures of Bacteriology*. Several interesting books on applied entomology have appeared in the last few years, but they each deal with links between particular aspects of the discipline and other domains: entomological practice and ecological theory (Palladino, *Entomology*), pest-control and chemical warfare (Russell, *War and Nature*) and a cultural history of the concept of 'pest' (Jansen, *Schaedlinge*). None of the authors sought to write a general history of economic entomology. At present the only major works on specifically agricultural disciplines are Wieland, *Wir beherrschen*; Boehm, *Einfuehrung*; Comberg, *deutsche Tierzucht*; Russell, '*Like Engend'ring Like*'; Fitzgerald, *Business of Breeding*; Wood and Orel, *Genetic Prehistory*; and three studies of agricultural chemistry: Rossiter, *Emergence of Agricultural Science*, Schling-Brodersen, *Entwicklung*, and Jas, *Au Carrefour de la Chimie*.

Germany, though several interesting studies on Britain and the Netherlands.[23]

By far the largest literature of this kind concerns the United States and focusses upon the agricultural faculties of the land-grant colleges, the experiment stations and the U.S. Department of Agriculture.[24] Many of the latter studies, however, are centennial volumes whose aims are celebratory rather than analytical, and only rarely have these works explored the origins of 'science-oriented' and 'practice-oriented' approaches in agricultural science.[25] What the literature on the United States has done rather more successfully, however, is to identify the contribution of American agricultural education and research to the industrialisation of agriculture. As is well-known, the mode of practice which has come to dominate Western agriculture over the course of the 20th century – and which has been exported to developing countries since the 1940s – is intensive cultivation, including synthetic fertilisers, highly bred plant varieties, mechanisation and irrigation. Since the 1970s, however, critics have argued that this form of agriculture has been a mixed blessing.[26] While productivity has increased enormously, bringing lower prices to consumers, and back-breaking labour has been greatly reduced, the relatively small family-run farm has paid a high price. For hi-tech methods have been particularly attractive to large farmers (as well as

23 On German agricultural education see Boehm, 'Geschichte des Landwirtschaftsstudiums'; on the emergence and development of the German experiment stations see Finlay, *Science, Practice & Politics* and Jas, *Au Carrefour de la Chimie*. I will discuss the literature on British and Dutch institutions in Chapter Six.

24 The literature on various American institutions is listed in Rossiter, *List of References*. The only available overview of the institutional development of American agricultural sciences is Rossiter, 'Organization of the agricultural sciences'.

25 An exception is Storer, *Science and Scientists*. Although he identifies two contrasting approaches toward research among experiment station scientists, however, Storer attempts to assess only 'attitudes' without examining their actual research activity. Moreover, the book's focus is very much restricted to individual differences, and since Storer examines but a single station, he is unable to say anything about systematic institutional variation.

26 For a discussion of the relevant literature, see Chapter Six.

to agricultural industry) who have been much better placed to benefit from them than have small farmers.[27] As we shall see, in Germany at least, such hi-tech methods emerged from the 'science-oriented' colleges while the 'practice-oriented' ones were more concerned to increase the productivity of small farms. As a result, 'academic drift' in agricultural education has been one of the driving forces behind the industrialisation of agriculture. An analysis of the processes whereby colleges became 'science-' or 'practice-oriented', therefore, can enlarge our understanding of the overall transformation of agriculture since 1900 as well as giving us a concrete sense of 'the road not taken'.

Agricultural higher education in Germany ca. 1900 lends itself especially well to a study of this kind for a number of reasons. To begin with, agricultural academics' social location within higher education was undeniably marginal. On the one hand, academics in these fields were fully aware that they were regarded as second-class citizens by colleagues in the sciences and humanities.[28] And although this situation was not peculiar to Germany, it was probably more acute there because of the especially intense anti-utilitarian ideology which dominated late 19th century German universities (from which for-

27 Indeed, agricultural scientists themselves believe their work is more apt to benefit large farmers (Busch and Lacy, *Science, Agriculture*, chp. 8).

28 On his appointment as professor of agronomy at the University of Koenigsberg in 1906, E.A. Mitscherlich reckoned that his cautious reception by colleagues was due to the fact that agriculture was not yet 'regarded as a fully fledged academic discipline' (Mitscherlich, *Lebenserinnerungen*, 18). At Bonn in the 1880s the agricultural botanist, F.A. Kornicke, found that mainstream botanists regarded the study of cultivated plants with condescension (Ullrich, 'F.A. Koernicke', 93). In Munich during the 1920s newspaper articles often suggested that the Agricultural Faculty was the poor relation (*Aschenbroedel*) (minutes of meeting of sub-committee, 13 Oct 1925, MK19682, BayStA). At Berlin in 1934 University staff felt humiliated when the city's veterinary and agricultural colleges were incorporated into the University as a separate faculty (Baumgarten, *Vom Gelehrten zum Wissenschaftler*, 90). See also Krzymowski, *Philosophie der Landwirtschaftslehre*, 39–46; von Ruemker, *Landwirtschaft u. Wissenschaft*; Schroeder, 'Hochschule oder Universitaetsinstitut?'; Roemer, *Beobachtungen auf dem Gebiete*, 111–112.

estry, veterinary medicine and chemistry also suffered).[29] Moreover agriculture's status-deficit was not helped by the relatively modest academic qualifications of both its staff and students.[30] In response, predictably enough, some professors of agriculture sought to fend off such criticism by playing down the practical side of their discipline.[31] But staff also ran into criticism from outside academe. Farmers, for example, routinely dismissed agricultural science as 'arid theory' (graue Theorie).[32] And much to the chagrin of academics, most farm-

29 On agricultural science's status problems in other countries, see Howarth, 'Science education', 365; Brierley, 'Some viewpoints of an applied biologist', 352–354; Harlan, *One Man's Life with Barley*, 116; Rossiter, 'Organization of the agricultural sciences'; Jacob, 'Reflections of an applied biologist'; Morrell, *Science at Oxford*, chp. 4; Joravsky, *Lysenko Affair*, 109, 188; Ede, 'School of agriculture'. For the status problem faced by Venezuelan agricultural institutions in the 1940s, see Vessuri, 'Foreign scientists'. On the ideology of the late 19th century German university see Ringer, *Decline of the German Mandarins*.

30 At Giessen for example the proportion of staff appointed without the second dissertation (*Habilitation*) in the few decades before the Second World War was somewhat less than 20% overall but was over 40% in agriculture (and even higher in veterinary medicine). And while about 10% of Giessen's professors overall were appointed while holding non-academic positions, the proportion for professors of agriculture was 30% (Moraw, 'Organisation und Lehrkoerper'). As far as student qualifications were concerned, on the eve of the First World War about half of the students in agriculture had no full secondary school diploma (*Abitur*) while the proportion among students of the humanities and natural sciences was two percent or less. Notable exceptions among the sciences were chemistry (15%) and pharmacy (80%), both of which also suffered a status-deficit (Titze, *Datenhandbuch*, 218–220).

31 While complaining, for example, that it was unfair to look down upon agricultural staff just because they addressed questions of practice, the professor of farm management, Richard Kryzymowski, acknowledged that some agricultural scientists were excessively utilitarian in their outlook (*rohe Praktiker*) and that the discipline must therefore develop a 'purer and more noble form' of science (Kryzymowski, *Philosophie der Landwirtschaftslehre*, 39–46).

32 Farmers' scepticism of agricultural science proved to be an obstacle to the establishment of the earliest German experiment stations in the 1850s and 1860s (Finlay, *Science,Practice and Politics*, chp. 2).On 'arid theory' see among others Muenzinger, *Aus meinem Leben*, 82; von Ruemker, *Landwirt. Versuchs- u Unterrichtswesen*, 17; Stutzer and Gisevius, *Der Wettbewerb*, 106; Wiedemann, 'Wissenschaft und Praxis'. I presume the phrase 'graue Theorie'

ers were also sceptical about the value of higher agricultural education.[33]

Academics were acutely aware of this tension between the worlds of scholarship and practical agriculture, and many sought to ameliorate it by declaring their goal to be the 'unity of theory and practice':

> In order that [...] the students are constantly in contact with the world of practice, and so that they do not become 'latin-spouting farmers' ['lateinische Landwirte'] or 'theoreticians', we make every effort between April and August to arrange weekly student-excursions to nearby farms.[34]

Numerous textbooks bore the subtitle 'based on scientific and practical foundations'; obituaries of agricultural scientists praised their subjects for combining both 'science' (or 'theory') with 'practice'; and an early journal of agricultural chemistry took as its motto 'practice with science'. The great strides which German agriculture had made since Liebig, the agronomist Kurt von Ruemker declared, were due to the 'intensive collaboration between theory and practice'.[35] To

was taken from Goethe's *Faust*, part 1 (Nr. 2038–2039) in which Mephisto says 'Grau, teurer Freund, ist alle Theorie, und gruen des Lebens goldener Baum' (All theory, dear friend, is gray while life's golden tree is green).
Once again, such scepticism was not confined to Germany. On American and British farmers' criticisms of 'book farming' see Fitzgerald, *Business of Breeding*, 135; Cooke, 'From science to practice'; Holmes, 'Science and the farmer'; Heape, *Breeding Industry*, 22; Danbom, *Resisted Revolution*, 41, 70. Schools of brewing came in for similar criticism (Teich, *Bier*, 229; Haerd, *Machines are Frozen Spirit*, 214).

33 Farmers are said to have joked about the crop failures of 'classically educated students of agriculture' (*lateinische Landwirte*) (Hansen, *Das landwirtschaftliche Unterrichtswesen*, 1st ed. 89). See also Feulner, 'Landwirt. Berufsbildung'; Kluge, 'Das Hochschulstudium', 66; Wagner, 'Zur Frage des Studiums', 92; von Ruemker, *Die Ausbildung des praktischen Landwirts*; Zoerner, 'Reform'; Oldenburg, *Ausbau*, 18; von Ruemker, *Mein Leben*, 114; Vossler, *Die Entwicklung Hohenheims*, 21.

34 *Chronik der Universitaet Halle, 1916–1926*, 207.

35 Schindler's *Der Getreidebau* was one such textbook. For obituaries see Wunderlich, 'Fruwirth'; Roemer, 'Begleitwort'. The journal of agricultural chemistry was *Der Chemische Ackersmann* (Finlay, *Science, Practice and Politics*, 93–94). The quote is from Ruemker, 'Entwicklung der landwirt.

some extent, of course, this was little more than rhetoric, but the interesting fact, as we shall see in subsequent chapters, is that some agricultural colleges made much greater efforts to bridge this gap than others.

Lastly, Germany from the late 19th century is a particularly convenient time and place to study this phenomenon because of the sheer diversity of its higher agricultural institutions (cf. Chapter Two). In addition to several independent colleges, agriculture was taught in the arts and sciences faculties of a number of universities as well as in one engineering college. Nor were the colleges identically situated or organised. Some were in cities while others were in rural areas. Some were only entitled to award a short and rather vocational degree while others offered the doctorate, and so forth. All of this institutional diversity offers rich material for comparative analysis, and in the following chapters I have taken advantage of it in order to establish how it is that some of these institutions evolved along a 'science-oriented' path while others grew increasingly 'practice-oriented'.

The book in outline

The next chapter ('Mapping the Institutional Landscape') provides orientation by introducing the reader to the wide range of variation among institutions of agricultural education (including their structure, funding, staff, students, curriculum, etc) as well as their geographical settings. It also surveys the variety of institutions to which they were linked and which exerted some influence upon their fates, including not only other academic institutions but also political and economic

Pflanzenzucht', 337. In a similar vein see Ruemker, 'Das landwirt. Versuchs- u. Bildungswesen', 107; Diener, 'Zur Geschichte', 471. Nor was Germany peculiar in this respect; 'science with practice' was also the motto of Iowa Agricultural College (Overfield, *Science with Practice*, 8) and of the British Royal Agricultural Society (*Journal of the Royal Horticultural Society* , Vol. 24 (1900), 45).

ones. And lastly it outlines a model of institutional dynamics which informs the rest of the book.

Chapter Three ('Science vs. Practice in the Evolution of the Agricultural Curriculum, 1800–1934') turns to the colleges and charts the ways in which the curriculum evolved from the early 19th century. It is a story of how and why the educational pendulum has swung back and forth between 'science' and 'practice'. It also serves to acquaint the reader with the various forms in which agricultural higher education was institutionalised and provides a map of the changing structure of the higher education system. I devote particular attention to the vigorous debates from the 1890s and again following the First World War over the relative merits of the colleges versus the university departments of agriculture since these debates often hinged on the question of 'science' versus 'practice'.

Instead of surveying the system as a whole, Chapter Four ('Exemplars: Science and Practice at Five Institutions') focuses in detail upon illustrative institutions over the period, 1890–1914. For each institution I provide a rudimentary sketch of its history and characterise each place's overall orientation toward science/practice, as evident from its teaching, research, staff and students. I show that while Berlin and Halle were relatively 'science-oriented', Hohenheim, Bonn and Breslau displayed a much stronger practical orientation. On the basis of comparative analysis, I conclude that the most important causes of an institution's orientation toward science/practice were its place on the academic status hierarchy, the economic interest-groups from which it could claim support, and political action by the relevant state government.

Chapter Five ('Manoeuvring for Position: Bavaria's Agricultural Institutions, 1872–1934') Building on the results of the previous chapter, I focus in more detail on Bavaria, tracing the evolution of the state's two higher education institutions – which contrasted in their orientations toward science/practice – as they competed with one another throughout the period. By following their histories over this longer period, I can adjust the model so that it is less a static sociological theory than a dynamic historical one.

Finally in Chapter Six I explore the model's more general applicability. I consider how it might illuminate the history of agricultural sciences and education in the United States and Britain, but more importantly, I argue that my findings are potentially relevant to a wide range of practical sciences, including engineering, medicine and management sciences. Finally, I return to the educational policy issue with which this chapter opened – the problem of academic drift – and consider what the model developed here might have to offer those concerned with the reform of higher technical education.

Chapter 2: Mapping the Institutional Landscape

When we stand on a hilltop to observe the panorama below and beyond, our eye takes in the variety of geographical features as it sweeps across the landscape. This chapter provides a similar panorama. I provide an overview of the diverse forms taken by German agricultural colleges: in their geographical settings, their formal organisation, the demonstration facilities they possessed, the structure of their teaching programmes, the orientation of their research activity, and the backgrounds of their staff. These are the 'species', as it were, which populated the institutional landscape. If we are to understand the origins of their variation, however, we will also need to look beyond them in order to survey the diverse 'habitats' which they occupied.

One such habitat consisted of the network of other higher education institutions. For the actions undertaken by each institution were often predicated on what the others were doing (and especially those in the same region since they were most likely to be competitors). Another habitat was the institution's actual geographical location since that opened up some developmental strategies while hindering others. A third habitat was political in character. Since higher education was financed, not by the federal government but by the states, institutions were accountable to one or other state ministry. And the states' involvement in agriculture was not restricted to higher education. In addition to secondary agricultural education (which provided employment for many graduates of the colleges), a variety of experiment and testing-stations dotted the landscape whose relations to education were also sometimes significant. Finally, of course, a crucial habitat was economic. Farmers and agricultural industry, as the most obvious constituencies for agricultural science and education, took an interest in the activities of the colleges in their vicinity. And as we shall see, the nature (and influence) of both of these user-groups varied greatly from one region to the next. Just as in ecology a species is affected by its habitat, so also in higher education; institutions were shaped in part through the academic, geographic, political and economic habitats which they occupied.[1]

A main task in this chapter, therefore, is to map the lay of this land. After charting the diversity of institutions in the first section, we turn to the immediate academic context: the structure and operation of the higher education system as a whole. Section three surveys the diversity of geographical settings before turning to the political and economic organisations with which colleges had to reckon. Having sketched the relevant terrain, in the final section I outline the model which informs the subsequent chapters. It attempts to integrate all of these habitats in order to explain why educational institutions differed in their orientations to 'science' and 'practice'.

The variety of institutional forms

Perhaps the most obvious difference among the institutions ca. 1914 was size. The college at Berlin was huge with 18 full professors (and institutes) while tiny Jena had but one. Student numbers also varied greatly. In 1911–12 for example Kiel, the smallest Prussian institution, had a grand total of two students while Halle and Berlin each had over 400. By 1926 Kiel had greatly boosted its student numbers to 116, but Halle still had over 600 and Berlin over 700. State expenditure varied accordingly; in Prussia in 1910, for example, Berlin received 450,000 marks from the state government while Kiel got less than 4,000.[2]

1 For an early use of ecological models in thinking about scientific institutions, see Rosenberg, 'Toward an ecology of knowledge'.

2 On Berlin vs. Jena see Schroeder, *Studium der Landwirtschaft*. On student numbers see Schroeder, ibid. and Hansen, *Das Landwirt. Unterrichtswesen* (2nd ed. 1920), 19. On expenditure see Oldenburg, *Der Ausbau*, 110.

Table 2.1 – Higher Agricultural Education Institutions in Germany ca. 1900

	Location (province)	*Ministry responsible*
Academies/colleges:		
Hohenheim	Wuerttemberg	Education, Wuerttemberg
Bonn	Rheinland	Agriculture, Prussia
Weihenstephan	Bavaria	Education, Bavaria
Berlin	Brandenburg	Agriculture, Prussia
Engineering college:		
Munich	Bavaria	Education, Bavaria
Universities:		
Breslau	Silesia	Education, Prussia
Giessen	Hessen	Education, Prussia
Goettingen	Hannover	Education, Prussia
Halle	Prussian Saxony	Education, Prussia
Jena	Thuringia	Interior, Thuringia
Kiel	Schleswig-Holstein	Education, Prussia
Koenigsberg	East Prussia	Education, Prussia
Leipzig	Kingdom of Saxony	Education, Saxony

As Table 2.1 indicates, the institutions also differed in their formal organisation. The 'academies' (or 'colleges'[3]) were independent institutions while agriculture at Munich constituted a small Faculty, and the universities had departments of agriculture which were all located in Philosophical Faculties.[4] The latter locations, as we shall see, restricted the autonomy of agricultural staff in some areas such as making appointments or awarding doctorates. In addition, the form of governance differed in academies versus universities. The universities,

3 For convenience henceforth I will refer to all of these independent institutions as 'colleges', though as we shall see below, the formal status of an 'academy' and a 'college' (*Hochschule*) were significantly different.

4 Until after the First World War when at several universities these faculties were divided into those for arts and for sciences.

engineering colleges (Technische Hochschulen) and the College at Berlin were led by a 'rector' who was a member of the professoriate who had been elected by his colleagues for a period of one or two years. This structure gave staff considerable influence over academic policy, in keeping with the emphasis since the early 19th century university reforms upon academic freedom. In contrast, lower-ranking institutions which offerred technical training – such as the academies at Bonn, Hohenheim and Weihenstephan – were governed by a 'director' who was appointed for an extended period by the Ministry concerned, an arrangement which gave the Ministry more control of the institution. This structure also gave directors far more power than rectors since the former were not accountable to the professoriate. Academic freedom was evidently thought unnecessary in such technical institutions.

Agricultural institutions also enjoyed varying relationships to neighbouring universities. Berlin was loosely affiliated to the local university, and Bonn's students were officially registered in the University's Philosophical Faculty, but the colleges at Hohenheim and Weihenstephan had few such connections. Each of these arrangements conferred both limitations and opportunities. Because of their proximity to nearby universities, for example, Berlin, Munich and Bonn could offer students a wide variety of non-agricultural courses (and in the former cases the attractions of a big city). Weihenstephan and Hohenheim, on the other hand, enjoyed easier access to the farming community but could offer only a limited range of non-agricultural courses. Thus an institution's location and formal organisation had potentially important consequences for its orientation toward science/practice.

Institutions also possessed very different facilities for teaching and research. For example, all of the institutions had science laboratories, but they were very unequally equipped with facilities used for demonstrating agricultural practices in teaching or for testing them in research. Thus almost all of the colleges had their own model farm, but very few university departments did, a fact which prompted concern among some that academics at the latter were in danger of withdrawing into the splendid isolation of the ivory tower. And although all possessed experimental fields, these varied hugely in scale, varying from less than one hectar at Munich and Berlin to over 100 at Halle.[5] In addition, some colleges (eg, Hohenheim and Weihenstephan) benefitted from the existence nearby of state agricultural institutions which were affiliated to the college and thus offered additional facilities for practice-based teaching and research.

The teaching programmes at these institutions differed in several important respects: in the qualifications offered, the entry requirements for students, and the structure of the curriculum. Two kinds of qualification were offered, and the differences between them illustrate quite clearly the distinctions, both intellectual and social, which are often drawn today between what is called 'education' and 'training'. The more 'scientific' of these took three years and was of higher status.[6] Students who had completed the full nine years of secondary school and passed the (*Abitur*) examination (the standard prerequisite for university admission) were entitled to embark upon this course in which a relatively large amount of time was devoted to the basic sciences. Graduates generally took up jobs in teaching, as administrators in government or in private-sector agricultural organisations, or as agricultural scientists. The more practically oriented qualification[7], lasting two years, devoted more attention to agricultural subjects and less to the basic sciences. Students admitted to this programme had generally passed only an intermediary examination after six years of secondary school (*mittlere Reife*) which entitled them to study vocational subjects but not arts, sciences or most other university subjects. Graduates often went on to run their own farms or to manage large estates. Before the First World War this was the only qualification offered at Hohenheim and Weihenstephan while the university departments, Bonn and Berlin also offered the more 'scientific' de-

5 On splendid isolation see von Ruemker, 'Landwirt. Versuchs- u. Bildungswesen', 104, 107. On experimental fields see Oldenburg, *Das landwirt. Unterrichtswesen* and *Mentzel und von Lengerkes Schreibkalender* (1913).

6 Before 1914 this was called the '*Landwirtschaftliche Lehrpruefung*'; from 1922 in Prussia it was renamed the '*Landwirtschaftliche Diplom Pruefung*'.

7 Before 1914: 'Diplom Pruefung'; from 1922 in Prussia: 'Lehrgang fuer den praktischen Landwirt'.

gree. After the War, by which time Hohenheim and Weihenstephan had been upgraded, they too offered both qualifications.[8]

Institutions also varied in the extent to which they insisted upon practical work experience prior to admission. From 1877 Prussia had required two years' work experience for those wishing to enter the three-year course, though in practice the requirement was not enforced. The lack of a formal requirement made for substantial differences of emphasis among institutions before the First World War. Berlin, for example, stated that two years' experience was merely 'desirable' while Weihenstephan's brochure for students stated that 'it cannot be emphasised enough' how important it was to obtain several years of broad and thorough practical training prior to enrolling.[9] After the war there were calls for a standard two-year requirement, and this was finally formalised in the new degree regulations introduced in Prussia in 1922 which brought about a high degree of uniformity in this respect.[10]

Even within a given type of degree there were substantial curricular differences from one institution to another, with some more 'scientifically' oriented and others more practical. At some (eg, Halle) first year students were advised to take courses exclusively in basic sciences (and sometimes economics) as these were thought to provide a 'foundation' (*Grundlagenwissenschaften*) upon which agricultural subjects would later build. At others such as Breslau, however, the sciences were conceived as 'ancillary' or 'contributory' subjects (*Hilfswissenschaften*); in their first year students took both basic

8 Oldenburg, *Der Ausbau*; Schroeder, *Studium der Landwirtschaft*.

9 On the 1877 statute see Boehm, 'Geschichte des landwirt. Studiums'; on Berlin see Lexis, *Die Hochschulen* and *Bericht [...] Weihenstephan*, 1912/13, 226.

10 On calls for a two-year requirement see Flaig, 'Eine ausserordentliche'; Hansen, 'Zur Ausbildung des Landwirts'. For the Prussian statute see Anon., *Pruefungsordnung fuer Studierende*. Even after 1922, some students outside Prussia were still exempted from the practice-requirement (eg, at Jena; see Schoeder, *Studium der Landwirtschaft*). The situation in engineering was similar. The Association of German Engineers (*Verein Deutscher Ingenieure*) had called in the 1880s for shop training prior to admission, but very few engineering colleges had made this compulsory before 1918 (Gispen, *New Profession*, 152; Albrecht, *Technische Bildung*, 354–355).

sciences and agricultural subjects. Here the aim was to familiarise students right from the start with agricultural problems so that they were equipped to gain maximal benefit from the basic sciences.[11]

These differences in the character of degree-programmes are useful indications of the quite distinct ways in which staff at various institutions conceived of agricultural education. But one must recognise that the significance of degree-programmes within the overall teaching activity was minimal. Since the average student only attended for about three semesters, the proportion of students who actually completed a degree was very small (much to the annoyance of the professors), varying from 1–2 % at Kiel or Koenigsberg to 17% at Munich.[12] Rather more important for the students was the *way* in which courses were taught. For although agriculture students everywhere took courses in the basic sciences, at the universities those courses necessarily addressed only general problems since they were taught by staff from the science departments and had to serve as well for science (or medical) students. The colleges, on the other hand, had their own science staff who were expected to tailor their courses to the concerns of agriculture students.

Another way in which a practical orientation was expressed in the curriculum concerns the teaching of farm management (landwirtschaftliche Betriebslehre). While all of the colleges had dedicated posts for this subject by 1910 (some of them already by the 1880s),

11 Schroder, *Studium der Landwirtschaft*; Boehm, 'Geschichte des Landwirtschaftsstudiums', 10. The same conceptual distinction was drawn in engineering education from the mid-1890s when a vigourous controversy broke out among academics over the proper role of mathematics in the curriculum (Hashagen, 'Der Mathematiker'; Hensel, 'Auseinandersetzungen').

12 Oldenburg, *Ausbau*, 16; Martiny, 'Zur Frage'; Hillmann, 'Das wissenschaftliche Studium', 122. For the figures for 1899–1911 see Hansen, *Das landwirt. Unterrichtswesen*, 2nd ed., 100. Kurt von Ruemker warned students against coming for just a semester or two, recommending a minimum of four (von Ruemker), 'Programm fuer das Studium', 8. (This phenomenon was not confined to agriculture. Students of engineering in the latter 19th century did not bother taking exams because industry attached little importance to degrees, and the situation in business schools was similar (Koenig, *Kuenstler und Strichezieher*, 85; Franz, *Zwischen Markt und Profession*, 62).

most of the universities did not (though it was taught by one or other member of staff virtually everywhere).

After the First World War most universities did acquire posts specifically dedicated to farm management, but even then the subject was assigned more importance at some places than others. At Bonn in the 1920s, for example, students did not encounter the subject until their second year while at Breslau they were obliged to take it in all three years of their course.[13]

Given such different entry requirements and curricular emphases, it is not surprising that the educational backgrounds of student bodies varied considerably from one institution to another. In the late 1920s, for example, the proportion of students who had gained the full 9-year secondary school qualification (*Abitur*) ranged from only 30% at Weihenstephan to 85% at Munich. Family backgrounds were similarly varied. At some institutions (eg, Goettingen) the proportion of students from farm backgrounds was three times as high as at others; moreover, the proportions of students whose fathers owned *large* farms also varied from place to place (8–9% at Breslau and Koenigsberg vs. about 30% at Jena, Munich and Goettingen). All of these differences in background, of course, also meant that students' career intentions varied substantially from one institution to the next.[14]

Identifying a 'scientific' vs. 'practical' tendency in an institution's curriculum or demonstration facilities is thus relatively straightforward. Attempting to characterise its research work, however, is another matter, not least because of the enormous diversity of disciplines constituting the agricultural sciences. Nevertheless, a few rather general features of an institution's research activity can tell us something about its orientation toward science and practice. One clue, for example, can be found in the types of publication in which its staff tended to publish. Journal outlets ranged widely from the popular to

13 On dedicated posts see *Mentzel und von Lengerkes Schreibkalender*, 1902–1930; Schroeder, *Studium der Landwirtschaft*; Titze, *Datenhandbuch*; *Landwirtschaftliche Jahrbuecher* for 1903 and 1910; Oldenburg, *Landwirtschaftliche Unterrichtswesen*. On Bonn vs. Breslau see Schroeder, *Studium der Landwirtschaft*.

14 *Deutsche Hochschulstatistik*, 1928. The proportion of students at Goettingen and Jena who intended to enter farming, for example, was relatively high.

the strictly academic. At one extreme there were the national agricultural newspapers aimed at farmers and other general readers (eg, *Deutsche Landwirtschaftliche Presse*, *Illustrierte Landwirtschaftliche Zeitung*) as well as the regional agricultural newspapers (*Landwirtschaftliche Wochenblaetter*) published in Bavaria, Wuerttemberg, Saxony and elsewhere. At the other end were journals in the basic sciences (eg, *Berichte der Deutschen Botanischen Gesellschaft* in botany or the *Zeitschrift feur induktive Abstammungs- und Vererbungslehre* in genetics). In between was a wide range of technical journals probably read by government agricultural officials, technical personnel in agribusiness, and academics (eg, *Journal fuer Landwirtschaft*, *Jahrbuch der Deutschen Landwirtschafts-Gesellschaft*, *Fuehlings Landwirtschaftliche Zeitung*). The choice where to publish, therefore, is an indication of which audience an academic wished to reach.

Although I have referred to an institution's 'research', this term is actually too narrow. For what we really need to focus upon are the activities – including publication though not restricted to it – through which staff attempted to make their expertise available to wider audiences. This means that we must also take note of staff's involvement in, for example, advisory activity. In some countries such as the United States, 'extension' work – whose aim was to bridge the gap between academe and farmers – was a responsibility of the colleges. In Germany, however, it was largely organised by non-academic bodies (see section three below), and some German academics were concerned that this arrangement fostered an unfortunate separation of science from practice.[15] Nonetheless, it is noticeable that some institutions made a greater attempt to maintain close ties to the farming community than did others. Before the First World War, for example, some institutions (eg, Breslau, Halle, Berlin) contributed to short courses for farmers during the winter (organised by the regional chambers of agriculture) while others (eg, Bonn) did not.[16] In addition, individual farmers were not the only user-group seeking advice from

15 For example, this was a view expressed by F. Aereboe to C.B. Hutchison on the latter's visit to Germany in the mid-1920s (box 1, file 33, Hutchison papers).

16 *Landwirtschaftliche Jahrbuecher* for 1903 and 1913.

academics; commodity associations, fertiliser companies, the sugar industry and commercial breeders were as well. And as we shall see in Chapters Four and Five, different institutions tended to lend their advice to different clienteles.

One can also learn quite a lot about an institution's research orientation from looking at the careers of individual staff. If one looks at the *curriculum vitae* of agricultural scientists in the decades around 1900, it is quickly apparent that unlike most disciplines in German universities at that time, there was no single standard career route. To be sure, some individuals did follow an educational trajectory little different from that taken by their contemporaries in the basic sciences: educated middle class family origins, nine years of secondary education at a classical grammar school (or after 1900 increasingly at a modern school), doctorate at a university, a further dissertation to qualify as a university lecturer (*Habilitation*), and then an academic appointment. But many others entered academe by completely different routes: eg, six years of secondary schooling at an agricultural school, higher education at an agricultural academy culminating in a non-research degree (*Diplom*), employment in some form of agriculture-related capacity – school-teaching, estate-management, working in an agricultural firm, administrator in an agricultural association or government agency, etc – before acquiring an academic appointment. And while career trajectories of this latter kind won little esteem in German university circles at that time, they were more likely to secure the respect of farmers and industrialists.

Consider some concrete examples: the agronomist, Eilhard Alfred Mitscherlich (1874–1956) and the specialist in farm management, Friedrich Aereboe (1865–1942), both of whom became eminent in their fields between the wars. Mitscherlich grew up in Berlin. His father was professor of surgery at the University while his mother's father owned several large estates. After graduating from a local classical grammar school, Mitscherlich wanted to study agriculture, but his father discouraged it, so the son divided his time at university between agriculture and physics. After receiving his doctorate in 1898, he went to the engineering college in Munich to work on soil physics. Three years later his Habilitation-dissertation was awarded a prize by the Bavarian Academy of Sciences, and he quickly took up a chair in agronomy at the University of Kiel (aged 27). Over the next few years he worked on the effect of the physical properties of soil upon yield, developing a new method for the measurement of the nitrogen content of soils, and on statistical methods to assess the outcome of field trials. By 1920 he had acquired the somewhat dubious reputation in agricultural circles of being one of the most 'theoretical' of agricultural scientists.[17] Thus Mitscherlich's social background, secondary education, and academic career were altogether typical of German universities of his time, and his work made little attempt to address the practical needs of farmers.

Aereboe's biography was quite different. Son of a school teacher and a peasant's daughter, he attended a classical grammar school in Riga (Latvia) but left before taking the advanced examinations (*Abitur*) in order to spend two years as an agricultural apprentice on small farms. At the age of 20 he attended an agricultural secondary school (*Ackerbauschule*) and worked for a while in several agricultural jobs before spending two years in the army. He then enrolled at the University of Jena, supporting himself with a variety of part-time jobs, and in 1892 received a degree in agriculture (*Diplom*) but was not permitted to stay on for a doctorate because he did not have the *Abitur*. Since Swiss universities were more flexible, he went to Basel where two years later his dissertation in plant physiology was accepted. Then followed a year of teaching at an agricultural secondary school, four years as head of the records division of the German Agricultural Society, and five years as the manager of a very large estate. Finally, in 1904 (aged 37) he was appointed to his first post as professor of farm management at the University of Breslau. His students have described him as free of the ivory-tower mentality then common in academe,[18] and one can see why. According to his memoirs, his textbook of agricultural policy addressed the problems of small farmers and was well received by them. As director of the college at Hohenheim from 1919 to 1921, similarly, he set up an advisory council of local farmers who exercised rights of veto over

17 Mitscherlich, *Lebenserinnerungen*; Mitscherlich, Lebenslauf, Sign = II, 1, Bd 5, TUM-Archiv.

18 Rolfes, 'Aereboe'; Ries, 'Zum hundertsten Geburtstag'.

staff on policy matters (cf. Chapter Four).[19] In his contributions to educational debate, as in his proposals at Hohenheim, he repeatedly argued that emphasis in the agricultural curriculum needed to be shifted away from general arts courses, as well as sciences such as geology and mineralogy, toward economic and practical matters. The task of higher education, he stressed, was not to train researchers but farmers and agricultural instructors.[20] Thus in contrast with Mitscherlich, Aereboe came from a relatively modest social background, received a vocational secondary education, and his higher education was fitted around employment of various kinds. Only after working in agriculture for ten years did he take up an academic career in which, not surprisingly, he stressed the importance of practical issues.

Having to recruit from a group of candidates whose careers were so diverse, appointments committees were confronted with a dilemma. Did one appoint candidates like Mitscherlich who had the conventional array of academic qualifications – but precious little practical experience – in order to allay the suspicions of agriculture among their colleagues in the arts and sciences? Or did one opt for those like Aereboe, who lacked certain standard academic qualifications but had a great deal of practical experience, with the aim of keeping farmers and the Minister happy?

The minutes of appointments committees can be illuminating on this issue. Any reader who has ever sat on such committees might be forgiven for doubting that their minutes would tell us much about what actually went on in the committee. But in fact the minutes of German committees during this period can be a remarkably rich source. Little attempt seems to have been made to disguise conflicts within the committees, and the arguments of individual committee members for and against various candidates are often recorded. Furthermore candidates were often assessed by the committees in terms of their 'scholarly' (*wissenschaftlich*) and 'practical' merits, so the minutes can be helpful in revealing an institution's perceptions of

19 On his textbook see Aereboe, *Jugendlust*, 163; on Hohenheim see ibid., 145–146.

20 Aereboe, *Allgemeine landwirt. Betriebslehre*, 639–650; Aereboe, *Jugendlust*, 150.

the ideal balance between theory and practice. In reading minutes, of course, one needs to distinguish between rhetoric and actual decisions. For example many committees declared their intention to nominate candidates who were strong on both the 'scientific' and the 'practical' side. This is hardly surprising, given that the minutes would be seen by the Minister in question (who was accountable to farmers) as well as – in the case of university appointments – by the Philosophical Faculty. But in practice the available candidates did not always match up to this ideal. Thus appointments committees were often forced to compromise, and the key question is *how* they did so.

Consider the appointment of a professor of animal-breeding at Giessen after the First World War. The appointments committee decided that it would only nominate candidates 'whose past work suggests that they will maintain the necessary ties with the farming community and that they will address breeding questions not just in a theoretical but also in a practical manner.' Finding that they had three candidates who seemed to offer both scientific and practical qualifications, the appointments committee ruled out one man because of his 'almost completely one-sided concentration upon purely scientific research problems'.[21] A few years later when Giessen was preparing to fill a chair in agronomy, one referee (who knew the circumstances at Giessen well since he had taught there until a few years previously) ranked a particular candidate very highly on the assumption that Giessen would 'want to cultivate ties to the countryside, on which the chambers of agriculture are usually so keen.'[22]

If institutions were in fact placing consistently different emphases upon science versus practice in making appointments, we would expect staff backgrounds to vary accordingly from one institution to another. Unfortunately systematic data of this kind are hard to come by. The well-known survey of German academics covering the Wilhelmian and Weimar periods undertaken by Helmut Plessner and colleagues, for example, did not record either the form of second-

21 Both quotes are from Bericht des Ausschusses fur die Vorbereitung der Berufung landwirt. Professoren, 24 Mar 1920, Sign = PrA, Phil Nr. 15 (Personalakte Hermann Kraemer), UA-Giessen.

22 A. Walther to Herr Geheimrat, 6 May 1925, Sign = Phil II k1, UA-Giessen.

ary schooling or the extent of non-academic work experience, both of which are of particular interest here. Furthermore the recently prepared electronic version of this database does not include the original data on agricultural scientists.[23] The only comparable survey I have found was undertaken by the Nazi regime during the mid-1930s (presumably to construct a register which could be used for surveillance). It consists of a file containing biographical information on approximately 6000 academics in the higher education system, and conveniently includes both of the variables omitted from the Plessner survey.[24] The utility of this database, however, is limited for a number of reasons. For one thing, it does not contain a substantial proportion of the full professors at various institutions. For another, given the growing similarity among higher agricultural institutions after the First World War (cf. Chapter Three), a snapshot of the academic community in the mid-30s is less likely to reveal stark differences among institutions than one from an earlier period. Furthermore such a 'slice' from a single period only captures very small numbers of staff at each institution so that statistical analysis is pointless. That said, the Nazi survey is useful on occasion, illustrating for example just how different the backgrounds of staff at some institutions could be, even as late as the 1930s. At Bonn, for example, all of the (seven) staff surveyed held doctorates; all but one of them were habilitated; and a majority of them had never worked outside academe. At Weihenstephan, on the other hand, one of the (seven) staff surveyed did not have a doctorate; none of them were habilitated; and all of them had worked in non-academic positions prior to their appointments.

When one examines all of the dimensions of each agricultural institution, perhaps the most important finding which emerges is that they tended to vary in concert rather than randomly. This means that many individual institutions can be characterised as tending toward either the 'scientific' or the 'practical' pole in their overall orientation

23 For the original survey by Plessner et al., see von Ferber, *Entwicklung des Lehrkoerpers*. On the electronic version see Ringer, 'Sociography'.

24 Nr. 10.000 – 10.025, R21, II. Anhang, Bundesarchiv Koblenz.

(see Chapters Four and Five). Having now surveyed the 'species' of interest, let's begin to explore the 'habitats' which they occupied.

The system of higher education as a habitat

One environment in which every academic institution found a niche – whether it sought one or not – was the network of other higher education institutions. To call this a 'network', however, is a bit misleading since its signal feature is that it was hierarchically structured. Institutions were ranked, both formally and informally, by status.[25] We know that such a hierarchy existed partly from looking at career trajectories; when academics moved from one institution to another, they generally left 'low' status ones in order to take up posts at 'higher' ones. Moreover this hierarchy was not static; ministers periodically exercised their power to upgrade engineering, agricultural or other types of vocational institutions (a process sometimes called 'academisation'). And one of the striking features of this process is the socially-charged language with which historical actors referred to it – 'increased value' (*Aufwerten*), 'equal entitlement' (*Gleichberechtigung*), 'of equal birth' (*ebenbuertig*) – all testifying to the academic status-hierarchy then prevailing.[26]

At the top of this hierarchy were the universities whose claim to status was justified by reference to their emphasis upon research and pure scholarship (*Wissenschaft*). Below them came those institutions who enjoyed the formal status of a college (*Hochschule*); while organisationally very similar to the universities, their weaker research traditions and focus upon applied scholarship was deemed the mark of lower status. Here one found the engineering colleges *(Technische*

25 Baumgarten, *Professoren und Universitaeten.*

26 Such social sensibilities were evident at the founding of the engineering college at Hannover when it was announced that the new college would be producing the 'scientific officers for the army of commerce' (Manegold, 'Entwicklung der TH Hannover', 42).

Hochschulen) which had acquired college-status in the 1870s and 1880s, and the veterinary colleges which acquired it in the late 1880s. At the bottom of the hierarchy were the 'academies' – of mining, forestry and agriculture – most of which acquired college-status only at the beginning of the 20th century. Here, too, were the youngest of the vocational institutions: the commercial colleges (*Handelshochschulen*) which were established between 1898 and 1920.[27]

The character of this hierarchy meant that among agricultural institutions Halle was at the top of the heap, as the oldest of the university institutes situated in one of the higher ranking universities whose research-orientation had given it a reputation among German universities already in the 18th century. One rung down came the institute at Breslau which lacked Halle's age and whose university ranked lower in status. In the middle of the ladder came Munich (by virtue of its location at an engineering college) and the Berlin Agricultural College since it was formally constituted as a college from its founding in 1881. At the bottom of the heap were the academies, among which Bonn arguably stood somewhat higher since its students were formally enrolled at the local University, and Weihenstephan was lowest since it was the last of the academies to acquire college-status and the right to award the doctorate.[28]

Given the status associations of 'pure' and 'applied' knowledge and the universities' location at the top of the ladder, one might be inclined to assume that the status hierarchy was nothing more than a statement of the obvious fact that some institutions were engaged in 'pure' scholarship while others were devoted to 'applied' subjects.[29] But status was about much more than purity; while 'purity' helped to

27 For dates of establishment and upgrading see Huber, *Deutsche Verfassungsgeschichte*, 944–947.

28 On the ranks of Halle and Breslau, see Baumgarten, *Professoren und Universitaeten*, 266 and 271. On the status differences routinely ascribed to universities versus the colleges, see Remy, 'Das hoehere landwirt. Unterrichtswesen'; Tschermak, *Leben und Wirken*, 163; Mitscherlich, 'Das Studium'. Symptomatic was the Association of German Universities' (*Hochschulverband*) reluctance to admit agricultural colleges to membership after 1918 (Nr. 12543, Rep 87B, GStaA).

29 I thank John Dettloff for prompting me to clarify this point.

confer status, it was by no means the only determinant thereof. This is evident from the career aspirations of scholars at that time. For generally speaking, whether they worked in 'pure' physics or in one or another 'applied' field, most academics preferred jobs in universities to those in engineering colleges (and even the latter were preferable to colleges of agriculture, veterinary medicine or forestry).[30]

Indeed, all of the organisational features which distinguished institutions at different levels of the hierarchy from each other – not simply 'purity' – were seen as significant indicators of status. Among these was the right to award the doctorate (acquired by the engineering colleges about 1900 and by colleges of agriculture after the First World War). This was important, not only for symbolic reasons but also because it gave the colleges leverage in requesting the provision of research facilities. Among those features which distinguished 'colleges' (as well as universities) from the lowly 'academies' were the possession of a 'rector' rather than a 'director' as the head of an institution, the right to award the status of *Privatdozent* to junior staff who had successfully completed the second dissertation (*Habilitation*), the right to nominate candidates for vacant posts to the minister of education, the right to demand of students the most advanced secondary school diploma (*Abitur*) for entry, the rights of staff to teach whatever they wished (*Lehrfreiheit*), and the right of graduates to qualify for jobs in the higher civil service. In organisational terms, too, each 'college' was also entitled to have multiple faculties (*Abteilungen*) and a senate as its highest academic decision-making body. Finally the ministry to which an institution was accountable was also important; the transfer in 1879 of the Prussian engineering colleges from the ministry of commerce to the ministry of education, for example, is reckoned to have given a major boost to their academic status.[31] (It bears saying that rights of this kind were not granted

30 For example, forestry staff at the agricultural college in Hohenheim were keen to transfer to the engineering college (chp. 4) while agricultural scientists at the engineering college in Munich petitioned the ministry to move them to the University of Munich (chp. 5).

31 On the characteristics of colleges see Manegold, 'Entwicklung der TH Hannover'. On the 1879 transfer see Manegold, 'Technology academised', 149; on post-1918 changes see Manegold, 'Geschichte der Technischen Hochschulen'.

overnight; in the case of the engineering colleges some rights were acquired in the 1870s while others were not conferred until after 1918.)

Each of these features counted as a badge of status, and invidious comparisons were constantly drawn on this basis. Since most university students were required to have the *Abitur*, for example, the possession of the *Abitur* by ones students was commonly taken to be a sign of status. Since many agriculture students at the universities did not possess this certificate, this fact was persistently used to discredit agriculture as a university discipline. (Chemistry and pharmacy also suffered for the same reason.) Conversely, those agricultural institutions with a relatively high proportion of students with the *Abitur* routinely broadcast this fact in order to elevate themselves within the hierarchy (as we shall see).[32] The academic qualifications possessed by an institution's staff were also deployed in this way. At the universities all staff were expected to have a doctorate (and often, the *Habilitation*); those academies which failed to measure up on this criterion were routinely derided.

'Academisation' is a ubiquitous feature of German technical education from the late 19th century, and a major aim of this book is to explain why it so commonly occurred among agricultural institutions. Although the concept has been used quite often (primarily in discussions of German engineering), however, it is seldom defined. Moreover, from the way in which historians have deployed the term, it is not clear whether it has an agreed meaning. For 'academisation' has sometimes been used to mean the social process whereby institutions were upgraded, at other times to denote the cognitive process whereby

32 Von Ruemker, 'Zur Frage der Ausgestaltung', 90. As late as 1928–29, for example, as many as 29% of agriculture students at the universities lacked the *Abitur* (and nearly 40% of those at the colleges lacked it): Ausschuss zur Untersuchung, 'Das landwirt. Bildungs- u. Beratungswesen', 46. The status-significance sometimes ascribed to the *Abitur* occasionally verged on the ludicrous. At Hohenheim, for example, while the professors of forestry were agreed that the quality of the forestry education there was superior to that at most universities, they argued nonetheless that Hohenheim's forestry institutes ought to be moved to the University of Tuebingen on the grounds that students with an *Abitur* (which Hohenheim's forestry students possessed) 'deserved' a university education (Franz, *Universitaet Hohenheim*, 79–80).

basic sciences (and other 'pure' subjects) came to play a more prominent role in the curriculum, and quite often to mean both of these things.[33] To use the term in the last of these senses might seem a dubious choice since in principle the social and the cognitive processes would appear to be analytically separable ones. Furthermore, as we shall see in the following chapters, they were also often separate in practice. It was perfectly possible, for example, for the state to endow an institution with improved science laboratories – or, on the other hand, demonstration facilities – without any effect upon their formal academic status. Similarly, although academic staff were nearly always keen on upgrading, that did not mean that the only institutional changes they ever favoured were those which strengthened the sciences.

Nevertheless, while these arguments are correct, it is important to recognise that as a matter of fact the social and the cognitive processes were often intimately linked. A number of historians have argued, for example, that while the growing centrality of the sciences to 19th century technology may well have arisen for more than one reason, a major cause of it, in Germany at least, was the social process whereby technical institutions sought to be upgraded.[34] One reason why these two processes were generally linked is that the legitimation given for social stratification was often a cognitive one. During the late 19th century, for example, ministers of education who tried to defend the higher status of the universities argued that science and technology were different forms of knowledge and thus required different insti-

33 In Manegold's 'Technology academised', for example, his use of the term sometimes implies a cognitve process (eg,142), at others a social one (146) and occasionally both (140). In Gispen's *New Profession* the term evidently means both cognitive and social change (73–75). Radkau's brief use of the term in his *Technik in Deutschland* appears to refer to both social and cognitive change (170) while Kocka (*Die Angestellten*, 94–95) uses it to designate just the social process.

34 Eg, Radkau, *Technik in Deutschland*, 160, 170; Manegold, 'Technology academised', 146. Although Koenig doesn't use the term 'academisation', he too argues that a main cause of 'scientisation' lay in technical institutions' growing organisational convergence upon the university as the normative model (Koenig, 'Stand und Aufgaben', 51–52; Koenig, 'Spezialisierung und Bildungsanspruch').

tutional provision. Accordingly, academics in the engineering colleges often concluded that in order to increase their status it was essential to introduce not only more basic science but also humanities subjects into the curriculum.[35]

But academisation also entailed a drift toward 'science' and away from 'practice' in a number of less obvious ways. For example, when institutions increased the entry requirement by demanding the *Abitur*, this meant allowing students with less practical experience to be admitted.[36] But this almost certainly had knock-on effects on the curriculum since staff could no longer take for granted the students' familiarity with practical agriculture when discussing the relevance of one or another aspect of the agricultural sciences. Similarly, when an institution acquired a rectorial constitution, this entailed a substantial degree of academic self-government, thus depriving the relevant ministry of an element of control over the running of the institution. Since the ministry was the usual channel through which economic interest-groups sought to exert leverage, this kind of organisational change enabled institutions to better evade pressures to move in a 'practical' direction (which is why, as we shall see in Chapter Four, such interest-groups sometimes opposed constitutional change). Awarding staff the right to teach whatever subjects they wished had the same consequence. Furthermore, in acquiring the right to award the doctorate, colleges had to adjust their appointments procedures accordingly. Henceforth increasing attention would be paid to candidates' research competence with an eye to their suitability as doctoral supervisors. And such competence was likely to be assessed by publication in specialised technical journals (rather than in farming magazines). Thus in all of these ways the result of academisation was to shift the focus in institutions away from 'practice' and closer to 'science'.

To say this, however, is not to claim that academisation necessarily had *negative* consequences for the technical sciences. To be

35 Manegold, 'Technology academised', 149–150; Gispen, *New Profession*, 73–74.

36 Throughout the late 19th and early 20th centuries those applicants lacking the advanced secondary school diploma (*Abitur)* were obliged to have at least one year of practical agricultural experience before taking up their studies.

sure, there is plenty of evidence that the heavy emphasis upon science or mathematics in engineering research and teaching often produced knowledge which users found pretty useless.[37] But it is important not to overlook the occasional counter-example. Thomas Hughes has pointed out that although the mathematics which 19th century German electrical engineers had to learn was of little use in inventing direct-current devices – Edison, for example, had little mathematical knowledge – it was very important in the construction and improvement of alternating-current equipment. Similarly, Edwin Layton has argued that the development of engineering theories which were sophisticated enough to help designers predict scale effects required the use of differential equations. Consequently most of the theorists of this 'dimensionless revolution' were European engineers who were well-versed in the calculus (as their American counterparts were not).[38] The situation in plant-breeding was comparable. Although Erwin Baur and Theodor Roemer were prominent advocates of the value of genetic theory as a guide to breeding practice, their breeding work was not thereby irrelevant (cf. Chapter Four). On the contrary, the new varieties developed in their institutes during the 1920s and 1930s were eagerly snapped up by commercial breeders. Thus there was no one right road to success in the technical sciences. And this suggests that those systems of higher education which managed to preserve diversity – embracing both scientifically- and practically-oriented institutions – might have been the locus of more innovation over the longer term.

37 Seeley, 'Scientific mystique'; Seeley, 'Research, engineering and science', 383; Dienel, *Ingenieure zwischen Hochschule und Industrie*, 553–564.

38 Hughes, *Networks of Power*, 141; Layton, 'Dimensional revolution'.

Situating institutions in their geographic, political, and economic habitats

As we have seen, each institution occupied a niche within the system of higher education, by virtue of which it was assigned an academic 'status'. If we wish to appreciate the full range of institutional diversity, however, we cannot stop there. For each institution also occupied a number of other 'habitats' – geographic, political and economic – which helped to confer upon each institution a distinct identity.

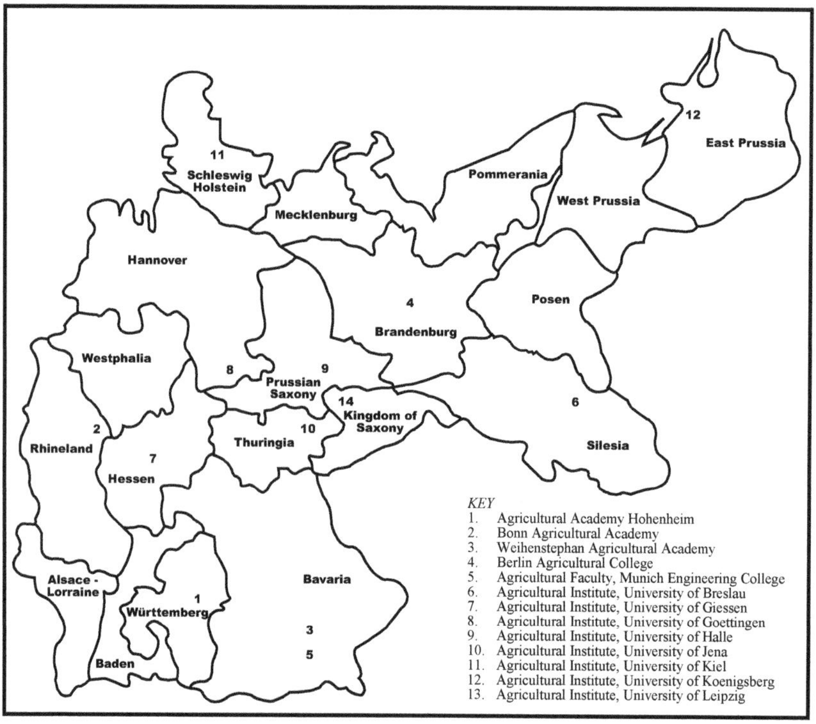

Figure 2.1 – Map of German states (1871–1918) showing Location of Agricultural Colleges

Consider the geographical setting, namely the physical features of an institution's immediate surroundings. One of the most striking differences among these institutions is the fact that some of them were located in large cities (Berlin, Munich, Breslau, Leipzig); others were in small rural towns (Hohenheim, Weihenstephan), and the rest were situated in a variety of university towns and smaller cities. Obviously each of these settings imposed certain constraints upon the institution but also conferred opportunities. Colleges in big cities, for example, were less likely to have substantial demonstration facilities nearby because of the cost of land, and it was harder for staff to stay in touch with the region's farmers, such that attempting to develop in a 'practical' direction was more difficult.[39] But an urban location offered potential advantages as well: access to one or another agricultural industry was sometimes easier, and recruiting students was relatively easy. Conversely, institutions located in rural areas were more likely to have excellent demonstration facilities, and it was much easier for staff to make contact with the farming community – making a 'practical' orientation much more likely – but on the other hand the lack of cultural amenities (and night-life) would deter certain students.

Moving from the immediate setting into the surrounding region, we find differences of climate, terrain, and soil type which made it easier to grow one crop rather than another. In this way geographical circumstances shaped the agricultural economy within which each institution had to function. For example the climate of Northern and Eastern Germany (eg, Pommerania, East Prussia, Brandenburg) is relatively dry with cold winters, and the soil is sandy. This made it suitable for growing rye or potatoes but not for wheat which requires richer soil and doesn't tolerate the cold as well. As a result the acreage devoted to rye in Brandenburg or Pommerania was eight or ten-fold that planted in wheat. In the richer soils of Silesia, Prussian Saxony or south Germany, on the other hand, the proportions planted in these

39 At Leipzig in 1890 the professor of agriculture called for degree programmes there to be shortened on the grounds that city life was in danger of derailing his students from pursuing a farming career (Boehm, 'Geschichte des Landwirtschaftsstudiums', 11).

two crops were two-to-one (or less).[40] Since sugar-beets, too, require good soils such as those in Prussian Saxony and lower Silesia, these regions became centres of the 19th century German sugar industry.[41] Thus staff at institutions in Koenigsberg (East Prussia) and Berlin (Brandenburg) had to reckon with the fact that rye was the major cereal crop in their region while key crops for those at Halle (Prussian Saxony) and Breslau (Silesia) were wheat and beets.[42]

Soil-type, of course, was not the only natural resource which varied from one region to another. In the early 1860s an estate-owner near Stassfurt (Prussian Saxony) noticed that a potassium-rich by-product of local salt-mining (potash) worked well as fertiliser. Within ten years there were 26 potash factories in the vicinity of Stassfurt, prompted in part by the high demand from regional sugar-beet growers, and by the First World War Germany had a world monopoly of potash production.[43] Unsurprisingly, the region's scientific institutions responded accordingly. In the late 19th century, for example, the experiment station in Halle had contracts with several fertiliser companies as well as with the regional sugar-growers association. Between the world wars similarly the professor of agronomy at the University of Halle (Theodor Roemer) was a consultant for the Potash Industry Syndicate (*Kali-Syndikat*), and one of his research fellows

40 *Statisches Jahrbuch fuer das Deutsche Reich*, Vol. 30 (1909), 60; on all crops see Engelbrecht, *Die Landbauzonen.*

41 Perkins, 'Organisation of German industry'.

42 Liebscher, 'Principles and methods', 356; Roemer, 'Entwicklungslinien'; Kryzymowski, 'Beziehungen zwischen Betriebsintensitaet'. Accordingly, some of Kurt von Ruemker's breeding work at Breslau used rye (von Ruemker, 'Korrelative Veraenderungen') and at Berlin one of Erwin Baur's major breeding goals, prompted by local estate-owners' desire to cash in on the increasing demand for wheat-breads, was to create a wheat variety which could thrive on the region's sandy soils (Baur,'Zukunftsaufgaben'). At Halle Roemer seems not to have bred rye, but about one-third of his most successful varieties during the 1920s and 1930s were wheats (Nathusius, *Theodor Roemer*, 31–32).

43 Welte, 'Bedeutung der mineralischen Duengung'; Borscheid, *Naturwissenschaft, Staat und Industrie*, 96–97.

was employed on a grant from the sugar-growers association (*Vereinigung mitteldeutsche Rohzuckerfabriken*).[44]

Terrain was similarly important. Insofar as hilly terrain is less suitable for mechanisation, such land lends itself better to pasture than to cultivation. Accordingly, patterns of land-use varied substantially from one region to another. In the relatively hilly regions of southern Germany (Bavaria, Baden, Wuerttemberg), for example, only about 40% of the land was under cultivation and 20% in pasture and meadow while in flat (and fertile) Prussian Saxony the proportions were 60% and 8% respectively.[45] As a result colleges in the southern states (Hohenheim, Weihenstephan and Munich) were obliged to devote particular attention in the curriculum and in their research to animal husbandry.[46]

Such diverse growing conditions meant that the nature of local agricultural problems, which institutions were expected to address from time to time, varied from one region to another. But perhaps more pressing than geographical setting was the institution's political habitat, by which I mean the network of state organisations which funded and regulated agricultural education, occasionally competed with it, and employed its graduates. Unlike the United States or Britain, Germany had no national department of agriculture until after the First World War (Reichsministerium fuer Landwirtschaft, est. 1920). This meant that agricultural matters (including education) were handled at state level and, depending on the state, were overseen in the main by either the ministry of agriculture or of education (cf. Table 2.1). The ministry was immensely important since it usually provided

44 Re: potash see 21 Jan 24, Dekan Schmidt to Kultus Ministerium, and 14 Jan 24, Roemer to Naturwiss. Fakultat, both in Akte c, PA 26005 (PA-Roemer), UA-Halle. Re: sugar see 11 July 23, Roemer to Kurator, Rep 33, Nr 63, UA-Halle.

45 *Statistisches Jahrbuch f.d. Deutsche Reich* (1909), 58.

46 For example, in 1917 Weihenstephan appointed Josef Spann as professor of animal-breeding, partly because he had worked as a veterinary inspector in the Allgau region of Bavaria, and his research was specifically directed at the problems of Alpine animal-husbandry (Personalakte Spann, AZ=II6, TUM-Archiv; MK43314, BayHStA). At the plant-breeding station in Weihenstephan, similarly, Hans Raum's job was to develop improved varieties of meadow-grasses and clover (Raum, 'Lebenserinnerungen').

virtually the whole of each institution's funding (apart from tuition fees).[47] It also appointed the directors of academies, appointed candidates to academic posts, and decided whether an institution ought to be upgraded. Since agricultural (and/or educational) policy could differ from one state to the next, this was another source of variability across the institutional landscape.

The other important state agencies were the research or testing stations. Experiment stations for particular fields (eg, plant-breeding, dairying, plant pathology, moor cultivation) were sometimes institutionally linked to nearby colleges, notably at Hohenheim and Weihenstephan. Where professors also doubled as station-directors, the stations' laboratory and field facilities became available for teaching, and the colleges gained a valuable asset. Testing stations (eg, for seeds, machinery) were often similarly linked, and this fact not only familiarised their academic directors with the problems encountered by local farmers but also earned the associated academic department a substantial income in fees from the tests. Proximity to these stations, however, could be a mixed blessing. Stations were generally better equipped for research than were academic departments (and arguably more tightly geared to the needs of practice).[48] Where an academic was working in the same field as a nearby experiment station, therefore, he had to choose his research topics carefully in order not to tread on toes.

Just as important as an institution's political environment, if not more so, was its economic environment. For ultimately the rationale for this kind of education was to further the development of the agricultural economy. By an institution's economic habitat I mean the major interest-groups in its region which represented either a section of the farming community or a branch of agricultural industry. Since the nature of the agricultural economy varied considerably from one

47 An additional source of research funding from 1920 was the National Research Council (*Notgemeinschaft der deutschen Wissenschaft*), one of whose committees was responsible for agriculture and forestry (Notgemeinschaft der deutschen Wissenschaft, R73, Bundesarchiv Koblenz).

48 On occasion ministries took this view (Min. Landwirtschaft, *Das landwirtschaftliche Bildungswesen*, 162).

region to another, each educational institution had to contend with a somewhat different combination of economic clienteles.

Let's begin with the farming community. With the availability from the mid-19th century of synthetic fertilisers and improved plant varieties, 'intensive' cultivation became possible, making for substantially increased yields. This entailed high levels of fertiliser, improved seed, weed control, mechanisation and sometimes drainage/irrigation. Fertilisers had emerged out of investigations by agricultural chemists, and following the expansion of both experiment stations and agricultural higher education in the latter 19th century,[49] other intensive methods became subject to improvement through systematic research in plant-breeding, plant pathology and agricultural engineering. Increasingly, therefore, 'science' was seen to be the key to such improvement.

Table 2.2 – Farm Size in various German Provinces (1907)[50]

	0-2 ha.	*2-5 ha.*	*5-20 ha.*	*20-100 ha.*	*>100 ha.*	*>500 ha.*
Prussia	62	15	17	5	0.55	–
Bavaria	36	24	34	6	0.08	15
Saxony	58	15	22	6	0.4	182
Wuerttemberg	53	26	18	2	0.04	1
Alsace	62	22	14	1.5	0.1	3
Mecklenburg	ca. 80	–	–	–	1.3	415
E. Prussia	53	–	–	11	1.5	403
W. Prussia	54	–	–	8	1.3	306
Brandenburg	58	–	–	8	0.8	467
Pommerania	59	–	–	7	1.5	708
Silesia	50	–	–	4	0.8	279
Rheinland	71	15	12	1.3	0.06	1
Germany total	59	17.5	18.5	4.5	0.4	3500

49 On agricultural experiment stations see Finlay, *Science, Practice and Politics*; on higher education see Chp. 3.

50 Data are from Oldenburg, *Ausbau des landwirtschaftlichen Unterrichts* (1920), appendix, pp. 104–105. All figures are numbers of farms as a percentage of the total in the relevant province, except for the final column which lists absolute numbers of farms larger than 500 ha.

But such methods were relatively expensive and thus out of reach for the large majority of German farmers whose farms were at most 5 hectars (12 acres) in size (Table 2.2).[51] These peasant farmers (*Bauern*) used manure for fertiliser, held back a proportion of their crop each season to use as seed for the next, and relied on horses or oxen for plowing (except for the few who gained access to machinery via a cooperative). For a small minority of landowners, however, whose farms exceeded 100 hectars in size (and were occasionally as large as several thousand), intensive methods were both affordable and attractive.[52] As one can see from Table 2.2, the proportion of these large farms (or 'estates') varied enormously from one region to another. In the flat northern and eastern provinces (Mecklenburg, East Prussia, West Prussia, Brandenburg, Pommerania and Silesia) the proportion of large farms was ten to fifteen times greater than in the hilly southern and western ones (Bavaria, Wuerttemberg, Alsace and the Rheinland). And this disparity is even more striking if one looks at absolute numbers. Each province in the North and East typically had several hundred farmers who owned 500 hectars or more while provinces in the South and West had but a handful. This meant that while all agricultural institutions were situated in regions where small farmers predominated, some were located in regions where there were also substantial numbers of estate-owners: at Halle, Koenigsberg, Berlin, and Breslau. At those institutions, therefore, staff could reckon with a significant minority within the farming community which was interested – as, by and large, the peasantry was not – in intensive cultivation and the science which seemed to drive it. Thus although the 'farming community' was an important clientele for all agri-

51 By comparison, the average British farm at this time varied between 30 and 40 hectars, depending on the region (Fletcher, 'The great depression', 40) while in the United States in 1900 the average farm was approximately 60 hectars (Danbom, *Resisted Revolution*, 3–4).

52 Various studies indicate that this is still true today. In the U.S., for example, large farms stand to profit more from introducing new technology, and the owners of large farms are also more likely both to adopt such technology and to be vocal in expressing their views on what public-sector agricultural researchers should be doing (eg, White, 'Interrelationships').

cultural institutions, its composition – and thus its needs – varied significantly from one region to another.

Farmers sought to advance their interests not just periodically in the polling booth but more often through agricultural organisations of various kinds. Among the earliest of these were the local and regional Agricultural Associations (*Landwirtschaftliche Vereine*).[53] The Associations dated from the beginning of the 19th century and grew very rapidly (from 15 in 1820 to over 800 by 1870). Though nominally voluntary associations, their establishment and funding in fact owed a great deal to state involvement since government officials saw them as a vehicle for promoting agricultural improvement among the peasantry as well as for obtaining information on the state of agriculture. (This mediating role of the Associations was formalised through the establishment in each state of an Agricultural Council (*Landwirtschaftsrat*) whose members were drawn from each of the local Associations and whose function was to advise the state government).[54] Peasants, however, at first constituted only a very small proportion of the membership (especially in the eastern states) which largely consisted of estate-owners, local and state agricultural officials, schoolteachers, clergymen and agricultural scientists. Over the course of the 19th century the proportion of peasants grew, but leading positions everywhere remained in the hands of land-owning nobles.[55] Nonetheless, in order to promote 'scientific' agriculture among the peasantry, the Associations conducted experiments, issued reports, held public lectures, organised exhibitions and demonstrations, and awarded prizes.

53 Von der Goltz, *Geschichte der Deutschen Landwirtschaft*, 370–372; Schling-Brodersen, *Entwicklung und Institutionalisierung*, 217–225; Ullmann, *Interessensverbaende*, 31–40.

54 These state Councils, in turn, sent representatives to the national German Agricultural Council (*Deutscher Landwirtschaftsrat*, est. 1872) which advised the Reich government. I have not discussed this and other national agricultural organisations since they did not figure in the regionally differentiated political landscape which is my concern here.

55 By the end of the century only a tiny proportion of the peasantry were members. Of the roughly five million German farmers ca. 1900 (the vast majority of whom were peasants), the total membership of the Associations numbered slightly over 200,000 (Altrock, 'Landwirt. Vereinswesen', 213).

They also set up model farms, provided financial support to local agricultural educational institutions, and took the initiative in the establishment of most agricultural experiment stations after mid-century.[56] What they did not do, however, was to address the most serious economic problem faced by peasants: the fact that despite early 19th century land-reform, they were still obliged to pay compensation-fees to the feudal landlords from whom they had supposedly been liberated. To address this would have been to enter the realm of the 'political' which would have been divisive within the Association and might well also have antagonised the regional government. Although officially recognised as the representatives of 'the farming community', therefore, the Associations more nearly reflected the views of estate-owners than of peasants.

During the revolutions of 1848, however, a new form of association took shape (which found rather less favour with governments). In opposition to the Agricultural Associations, peasants in many regions organised themselves into local Peasants Associations *(Bauernvereine)* which lobbied governments in order to assert their rights against large landowners. Although these Associations declined with the collapse of the 1848 Revolution, they were succeeded from the 1860s by a more successful movement which affiliated local to regional Peasant Associations (*Vereinigung der christlichen deutschen Bauernvereine*) and were particularly well-developed in Catholic parts of the South and West. Some of the regional Associations' activities were focussed upon advancing the economic interests of peasants, including the formation of cooperatives and adult education. Occasionally, too, these organisations spoke out on matters of higher education.

Peasant dissatisfaction with the Agricultural Associations did not go away,[57] and during the 1880s discussion began in government circles on how to get the peasantry more engaged in agricultural improvement. The rural unrest of the 1890s focused minds on this question, and various peasant organisations called for the replacement of the Agricultural Associations with an organisation which would

56 Finlay, *Science, Practice and Politics.*

57 Blackbourn, 'Peasants and politics'.

truly represent all of the farming community. In 1894 a new kind of organisation was duly created in Prussia (one in each province) and subsequently in other states: the Chambers of Agriculture *(Landwirtschafts-Kammer*). Though they did not replace the Agricultural Associations everywhere, the Chambers took over all of their property, several of their powers, and were designated as the official representatives of Prussian agriculture.[58] Most of the Chambers' members were elected by the farming community, giving it a more representative character than the Associations. Relatively well-funded (by a land-tax), the larger of the Chambers employed several hundred persons who administered divisions devoted to a variety of agricultural matters (eg, those for fertiliser, credit, cooperatives, animal husbandry, agronomy, and the press). Their education divisions put on short-courses (to which academics often contributed), offered economic and tax advice, employed peripatetic teachers (*Wanderlehrer*) who brought technical advice to farmers in isolated rural areas, and published weekly agricultural newspapers aimed at the peasantry (usually entitled *Landwirtschaftliche Wochenblaetter*). The Chambers also possessed a range of important testing and experiment stations, experimental fields and farms whose work was at the disposal of the region's farmers. Thus the Chambers, more heavily devoted to extension work than were most higher education institutions, provided the latter with a link to the farming community (where one was sought).[59] And unlike the Agricultural Associations, the Chambers were assigned not just a technical function – to advise government authorities and promote agricultural progress – but also a political one: to defend the interests of the region's farmers.[60] This meant that they did not hesitate to lobby

58 Achilles, *Deutsche Agrargeschichte;* Bergmann, *Bayerischer Bauernbund*, 80–88; von Ruemker, *Die Entwicklung der Landwirtschaft*; Altrock, 'Landwirtschaftskammern'; Altrock, 'Landwirtschaftliches Vereinswesen'.

59 In Hessen-Nassau, for example, the professor of agriculture at the University of Giessen obtained funding from the regional Chamber so that his Institute could operate several testing stations on the Chamber's behalf (Reinhold and Rolfes, 'Zur Geschichte der Landwirtschaftswissenschaft', 352).

60 Formally constituted by law (and funded in part by state governments), the Chambers might thus be equally regarded as part of each educational institution's 'political' as its 'economic' habitat. Since historically they represent a

higher education institutions whenever they felt that those interests were at stake.

Important though they were, however, Germany's five million farmers were not the only significant actors in the agricultural economy. For numerous branches of agriculture-related industry also took an active interest in their local higher education institutions, and academics had to try to take advantage of this fact. A case in point were the regional cattle-traders associations who funded improved facilities for animal breeding in Berlin and Halle after the First World War. Aware of this, the University of Giessen persuaded the Hessian ministry in 1919 to fund an additional chair of agriculture, though they realised that because of the weak economy, funding for the associated *institute* would have to come from private sources. Reckoning that such funding would be easiest to obtain from the cattle-traders, the appointments committee decided to dedicate the chair in question to animal-breeding.[61]

In the chapters to come we shall encounter comparable examples with respect to the brewing, sugar and flour-processing industries as well as fertiliser companies, commercial plant-breeders, and distilleries. In some instances these firms, too, lobbied the ministry on the establishment of new posts in favoured fields, but they also funded research and testing in academe, purchased commodities produced by academics, and lobbied for the appointment of candidates to academic posts. Here, then, was a final source of institutional variability.

Having surveyed the landscape of higher education, the scale of the problem should now be evident. For that landscape was dotted with a bewildering array of institutions, differing in many respects from each other. Furthermore each of them occupied a slightly different environment, defined by a distinctive combination of academic,

successor to the Agricultural Associations, however, I have arbitrarily treated them as part of the 'economic' sphere.

61 On Berlin see Nr. 20091, Rep 87B, GStaA. In 1914 the German Dairy Association petitioned the Ministry to establish a chair of dairy science at the College in Berlin (Dt. Milchwirtschaftlicher Verein to Min. Landwirtschaft, 16 Apr 1914, Nr. 20058, Rep 87B, GStaA). On Giessen see 'Bericht des Ausschusses fuer die Vorbereitung der Berufung landwirt. Professoren', 24 Mar 1920, Sign = PrA, Phil Nr 15 (Personalakte H. Kraemer) UA-Giessen.

geographical, political and economic habitats. This diversity of environments was potentially important, for as should now be clear, each local configuration offered the institution within it a set of constraints and resources which potentially shaped the way in which the institution developed. The central problem in the book, therefore, is to work out which of these numerous variables – be they structural features of the institutions themselves or one or other of their various habitats – were most important in channelling the institution along either a 'scientific' or a 'practical' path. In the next three chapters I use comparative analysis to try to answer this question, but it may be useful beforehand to outline my conclusions.

Understanding institutional dynamics

The model which informs the following chapters is loosely based upon Bourdieu's concept of the 'field'.[62] For our purposes it will be convenient to distinguish between the *academic field*, the *geographical circumstances*, and the wider *politico-economic field* in which each higher education institution had to operate (Figure 2.2).

62 For an introduction to the concept see Bourdieu, 'Field of cultural production' or Bourdieu, 'Specificity of the scientific field'. For his use of the concept in analysing higher education see Bourdieu, *Homo Academicus*.

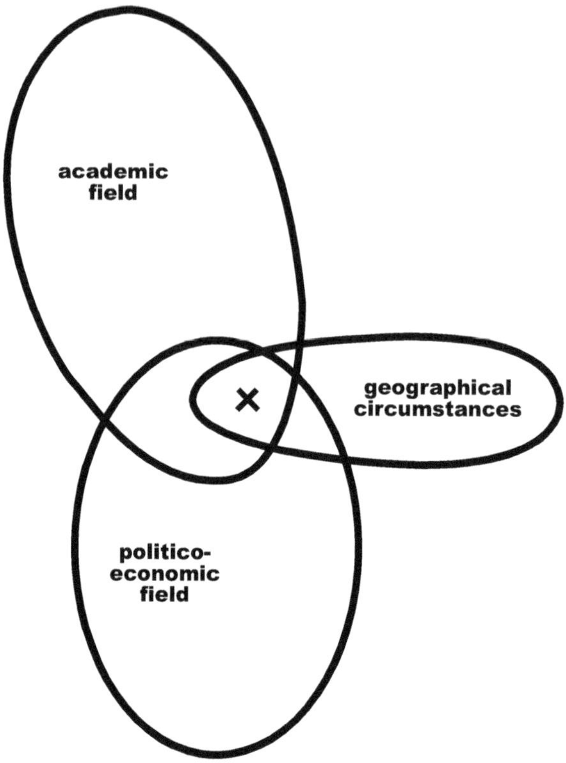

Figure 2.2 Diagram of an Institution (X) in its Academic Field, Politico-economic Field and Geographical Circumstances

The academic field is simply the constellation of other academic institutions making up the system of higher education. The geographical circumstances are the physical features of the institution's location. And the politico-economic field consists of the relevant state institutions as well as economic 'clienteles' or 'constituencies' in the region inhabited by an institution. The model assumes that each agricultural institution, endowed with certain strengths and weaknesses by virtue of its geographical circumstances, had to seek legitimacy, whether in the intangible form of 'approval' or in the material form of resources, from one or more of the agents in the academic and politico-economic fields. Thus every institution, taking into account its particular circumstances in all three domains, had to 'decide' what strategy to pursue in order to secure, wherever possible, both political acceptability and academic legitimacy. And whether an institution developed in a 'scientific' or a 'practical' direction depended on the strategy which it was able to pursue.

We have already seen how an institution's geographical location might affect the strategy it attempted to pursue. Now consider its position within the academic field. An institution's rank on the status hierarchy was something over which it had little control but which nonetheless had important consequences: for its recruitment of staff and students as well as for the legitimation strategies available to it. Institutions ranking high in the status hierarchy, for example, enjoyed a legitimacy in the eyes of the state, the academic community and sectors of the public. As a result, they rarely requested improved demonstration facilities or introduced more practical subjects into their curricula since (a) they had relatively little need of recognition from the politico-economic realm and (b) such changes would have compromised their status by reducing the differences between them and those institutions at the bottom of the ladder. For those at the bottom of the ladder, by contrast, there was little point in seeking to enhance academic status through mimicking the universities; this would have been at best a long-term strategy. The only realistic source of legitimacy available to such institutions over the medium-term was to try to ignore the academic status-hierarchy and concentrate instead upon expanding their utility with the aim of securing recognition from the politico-economic field. If continuity was the hallmark of institutions at the top and bottom of the hierarchy, for those in the middle ranks the strategy was to climb. Maintaining status if one is halfway up the ladder is a precarious business; for these institutions it was important to maximise their resemblance to the universities but imperative above all to maintain or enlarge the social distance between themselves and those at the bottom. Accordingly, it was here that staff lobbied hardest for academisation.

Thus an institution's position within the academic field predisposed it to develop along either practical or 'scientific' lines.

Whether it *succeeded* in this strategy, however, is another matter. In part this depended upon whether its geographical circumstances were favourable. But in addition, arguably all higher institutions must seek legitimacy, not only within the academic community, but also from the wider society. For agricultural institutions, of course, this kind of legitimacy was especially crucial, and to secure it required engaging with the politico-economic field.

Within this field, however, one might wonder whether the state (more properly, the ministry in question) should be treated as an independent agent. At first sight this looks plausible since, as we have seen, the ministry had very considerable say over the development of each institution. On the other hand, it is also clear that before taking the relevant decisions, ministerial officials routinely consulted with various interested agricultural organisations and were certainly not indifferent to political pressure. Thus one might with some justification conclude that the state is better conceived as a mediator of interest-groups in the agricultural economy rather than as an independent agent.

But to judge from the German case, neither of these interpretations is quite adequate. Although there is no doubt that in cases where powerful groups endorsed a particular policy, ministers were not able to ignore such pressure, there were also instances where the state was still obliged to act, even though the arguments of lobby-groups did not point clearly toward a particular policy. For example, what was to be done when:

- interest-groups are weak or indifferent? (In the 18th century, for example, state action established engineering schools and mining academies in Germany (as well as in France) in order to train personnel to run state enterprises. And in early 19th century Prussia the state's aim in educational reform – including the creation of secondary commercial and technical schools – was to promote industrialisation. That employers were not the driving force behind these reforms is evident from the fact that the level of instruction went well beyond the actual level of industrial production, so that for decades there were few jobs in industry which were appropriate for graduates).[63]

- two major interest groups were in conflict over policy? (eg, during the 1920s, as we shall see in Chapter Five, the main farming organisations in Bavaria were divided over the question of which of its colleges of agriculture should be closed).

- highly complex decisions must be taken, on which even the experts cannot agree? (eg, at the start of the 20th century what was the most effective way to increase the prosperity of peasant farmers: encourage the use of intensive cultivation methods or promote the formation of cooperatives so that peasants could buy and sell more efficiently?)

- different ministries are divided over the correct course of action? (eg, as we shall see in the next chapter, after the First World War the Agriculture and Education Ministries in Prussia could not agree on how best to improve agricultural education).[64]

Despite the conflict and ambiguity in all three of these situations, the ministry responsible must still act. And in the absence of clear signals from the agricultural economy, that decision will have to be based on the ministry's judgment. It is for this reason, therefore, that we need to think of the state as an occasionally autonomous agent capable of shaping educational institutions, even if in other instances it did little more than implement the demands of major interest groups.

63 Kocka, *Die Angestellten*; Manegold, 'Geschichte der Technischen Hochchulen'; Brose, *Politics of Technological Change*. By the end of the 19th century, to be sure, German industrialisation was growing apace, and economic interest groups were then well able to bring pressure to bear upon governments.

64 In 19th century France there were similar clashes over technical education between the Ministries of Public Instruction, of War and of Commerce (Day, *Education for the Industrial World*; Shinn, 'Reactionary technologists').

If the role of the state in forming institutions is complex, that of economic interest-groups is more straight-forward. Obviously the farming community's willingness to call for – or condemn – increased appropriations for educational institutions was crucial. And as we have seen, different sections of the community had different preferences in this respect, large farmers being rather more sympathetic to 'scientific' education than were peasants. As far as agricultural industry is concerned, it might seem as though firms would favour 'practically' oriented institutions, but industry's perspective appears to have been closer to that of large farmers.[65] As long as the agricultural economy was in good shape, firms could afford to invest in research of a more 'scientific' kind with an eye to the longer term. One sees this, for example, in the agronomical research of Eilhard Alfred Mitscherlich. His work at Koenigsberg on the nitrogen-content of soils, widely perceived during the 1920s to be 'theoretical', was initially supported by the German Agricultural Society, but the Society withdrew funding after a few years because the work was felt to be 'purely scientific'. For the next 35 years, however, this work was funded by the fertiliser industry who, according to Mitscherlich at least, never demanded that his work be directly applicable.[66]

So much for the main players in both the academic and politico-economic fields. How do they account for an institution's developmental path? The argument is a simple one: that development is shaped by a small number of countervailing forces. On the one hand, as we shall see, staff at most places sought the academisation of their institution (though the extent of the pressure varied from one place to another). Whether an institution actually succeeded in moving in the corresponding 'scientific' direction, however, depended upon its geographical circumstances and the composition of its politico-economic field. For such a development could be strengthened if the

65 The same was evidently true at this time in parts of Latin America where estate-owners and fruit companies saw an advantage in fundamental agricultural research (McCook, 'Promoting the practical'). On the ways in which corporations, trade-associations, and commodity groups attempt to shape the public-sector research agenda in the United States today, see Hadwiger, *Politics of Agricultural Research*.

66 Mitscherlich, *Lebenserinnerungen*, 22–33, 59–60.

dominant interest-groups in the region were sympathetic to 'science' but stymied if they were not.[67]

Let's consider, for example, how the model accounts for academisation. Occasionally authors have suggested that this process was inherent in the development of the technical sciences,[68] but there are two obvious problems here. For one thing, calling a process 'inherent' does not actually explain very much, whereas it is quite obvious that academic staff have often *acted* so as to further this process. Furthermore, it is clear that the process cannot be inherent in the development of these sciences because, though academisation is common, it has not invariably occurred. A case in point is German engineering education which began to shift in a practical direction from the 1890s,[69] and we shall see the same thing occurring at a few agricultural institutions around the time of the First World War. What requires explanation, therefore, is why the forces known to champion academisation do not always succeed. The answer is that agents in the politico-economic field – whether the ministry or particular economic interest-groups – have sometimes been actively opposed to upgrading. Thus academisation is not an automatic or 'inherent' process but rather a context-dependent one which arises through the interaction between relatively constant pressures in favour (from academic staff) and the regionally variable perceptions and powers of agents within the politico-economic field which may further or hinder it.

Having outlined a simplified version of the model above, however, let me point out a few complications. First, what is perhaps not

67 Because my concern at this point is with the causes of academisation, I have here singled out academic staff as active agents while state and economic institutions are portrayed as responsive. It is, however, emphatically not the case that the latter routinely took a passive stance in relation to higher agricultural education. As we shall see in the following chapters, on numerous occasions both ministries and economic interest-groups were quick to take initiative and apply pressure in order to shape the colleges in desired ways.

68 Haenseroth and Mauersberger, 'Das Dresdener Konzept'; see also the sources cited in Gispen, *New Profession*, 75, fn 32.

69 For just one example among many, see the discussion of German electrical engineering education in Koenig , 'Science-based industry'. Something similar happened at the Ecole polytechnique from 1914.

obvious so far (but will become clear in Chapter Five) is that the model is a dynamic one; the statuses of competing academic institutions, the policies adopted by ministers, and the aims of economic actors all changed over time. With both academic and politico-economic fields in flux, this meant that the development strategies pursued at any one institution were necessarily also fluid. Second, it bears saying that the model is not a reductive one. Although some factors were more influential than others, no single factor was all-important. And as soon as one begins to analyse the effects of such factors, it becomes quickly apparent that their effects were intertwined and cannot be easily separated out. Consider what happens when a ministry decides to provide improved experimental grounds at an institution. Once in place, these facilities will have substantial knock-on consequences. Obviously a college well-equipped with such facilities will attract different kinds of staff and students than one with excellent science laboratories. And the presence of staff and students with particular interests, ambitions and academic qualifications will in turn tend to foster an ethos in the institution which reinforces the effects of those facilities. Moreover, on arrival new staff are likely to adapt their teaching and research so as to take advantage of their institution's particular strengths. This, in turn, may enable staff to acquire the trust of a particular local constituency, a trust which can in turn be converted into further resources.[70] Thus when a minister decides to endow two institutions with different kinds of facilities, this triggers a cascade of related processes, all of which contribute to the progressive divergence of these institutions' orientations toward science/practice. Out of this cascade, to be sure, we can identify the original ministerial decision as crucially important for the institution's subsequent development, but the subsequent processes are inextricably linked. We cannot, therefore, assess the importance of each factor simply by looking at it in isolation. Instead we need to think about

70 A case in point might be Adolf Muenzinger. Drawn to Hohenheim in the 1920s because of its strong ties to the farming community (which in Wuerttemberg was dominated by the peasantry), he focussed his research in farm management there upon the problems of small farmers in a way which generated practical results and favourable public attention (Muenzinger, *Aus Meinem Leben*).

each institution as a complex of *interacting* factors (ie, as a 'system') in order to understand how it evolved over time.

Having outlined the main features of the model, let us now put it to work. The next chapter will trace the longer term transformation of higher agricultural education during the 19th and early 20th centuries.

Chapter 3: Science versus Practice in the Evolution of the Agricultural Curriculum, 1800–1934

Since its beginnings in the 18th century, higher agricultural education in Germany underwent a number of fundamental changes over the next 150 years, in its institutional locus as well as in the general character of the curriculum. From its origins as a lecture-based subject in the administrative sciences at some 18th century universities, it was established from the early 19th century in the new agricultural 'academies' which were not only more practically oriented but also assigned a greater role to the natural sciences in the curriculum. By the 1860s, however, as the academies came under attack for being overly practical, new institutes of agriculture were founded at several universities with the explicit intention of developing the subject as an applied natural science. From the 1890s and again after the First World War this model of education itself came under fire for allegedly ignoring the economic and practical dimensions of the subject. Thus this history was effectively a tug of war between two broadly distinct conceptions of the nature of agricultural science whose proponents had different views of the proper balance between science and practice. Over the long term, however, the process of academisation gradually eroded the institutional diversity of the pre-First World War period, and science-oriented forms of education came to predominate.

From an administrative discipline to a natural-scientific one

A form of higher agricultural education existed in Germany since the early 18th century. For agriculture fell within the remit of the new university chairs of 'cameralism' (*Kameralwissenschaften*, sometimes translated as 'administrative sciences'), one of the earliest of which was established at the University of Halle in 1727.[1] Along with agriculture, these professors taught forestry, mining, architecture and civil engineering, accounting and finance, and law and administration, mainly to students who would subsequently become managers of state mines, forests and domains. Soon after these chairs were founded, the first German textbooks on agriculture appeared. By training, however, cameralist professors were usually lawyers, theologians or philosophers rather than natural scientists, and they rarely had any practical experience of agriculture. As a result their writings on the subject were often dismissed by experimentally inclined farmers as 'arid theory'. Toward the end of the century some cameralists were themselves calling for more specialised university institutes dedicated to agriculture where teaching would be of a more practical kind.[2] At about the same time such a chair was in fact created at the Karlsschule, an institution of higher education in Stuttgart whose aim, like that of its predecessors, was to train civil servants who would administer Wuerttemberg's domains.[3]

By the turn of the century, however, Albrecht Thaer had become convinced of the need for a different approach. Trained as a physician, he became familiar with British agriculture while serving as George

1 Von Ruemker, 'Gehoert die Landwirtschaftslehre?', 182.

2 Giessmann, 'Einfluss der Entwicklung'; Mueller and Klemm, *Im Dienste der Ceres*, 18ff.

3 Boehm, 'Geschichte des Landwirtschaftsstudiums'; Schwabe, 'Zur Idee'; Franz, *Universitaet Hohenheim*, 14; Lindenfeld, *Practical Imagination*; Becker and Schmitt, 'Entwicklung der Agrarwissenschaften'.

III's personal physician in the late 1790s.[4] Returning to Germany in 1798, he published a three-volume report on the most advanced agricultural methods he had observed. He also conducted his own agricultural experiments and continued to publish works on practical agriculture. In 1804 he took up a cameralist chair at the University of Berlin and gave lectures there on agriculture but soon came to the conclusion that the university environment was not conducive to the subject.[5] In 1806 accordingly, he established a private school of agriculture (on a 300 hectare estate at Moeglin outside Berlin) which was to become the model for nearly all German higher agricultural education into the 1860s. Unlike the cameralists, Thaer emphasised the necessity of both science *and* practice. On the one hand, pupils were expected to have spent several years in agricultural work prior to registering at his school, and the school had its own working farm whose problems the pupils were to tackle. But Thaer also insisted that pupils study relevant subjects: both basic sciences and agricultural ones such as agronomy, animal husbandry and – above all – farm management (*Gewerbelehre*). By 1858 eleven 'academies' of agriculture had been established in the various German states, modelled upon that at Moeglin and seeking to achieve the same close links between science and practice.

Although Thaer's academy became the most influential model for agricultural education during the first half of the century, the universities did not willingly surrender agriculture to this new home. One Bavarian proposal (which was never implemented) adopted Thaer's emphasis upon the integration of science with practice but wanted the course to be taught at a university.[6] At Goettingen the new academy's initial reliance upon the University for teaching of the basic sciences allowed University staff to keep practical education at bay, delivering a strongly science-oriented curriculum until the late 1850s on the grounds that the would-be estate-owners who studied

4 Mueller and Klemm, *Im Dienste der Ceres*, 39ff; Finlay, *Science, Practice and Politics*, 21ff.

5 Giessmann, 'Einfluss der Entwicklung'.

6 Endres, 'Zur Fruehgeschichte'.

there needed such an education to maintain their social position.[7] Although the University at Goettingen eventually ceded control to the academy, universities elsewhere managed to retain agricultural education. At Jena in 1826, for example, the professor of cameralism, Friedrich Gottlob Schulze, succeeded in establishing an institute for agriculture (privately endowed, it was never taken over by the University).[8] To be sure, Schulze accepted the emphasis upon practical experience in Thaer's model: he insisted that his students worked in agriculture prior to registering at the University, and attached to his institute was a small experimental farm. But he placed rather less emphasis than Thaer upon the natural sciences, and his main concern was that students should acquire the 'all-round learning' (*Allgemeinbildung*) which he believed only a university could provide. By 1860, however, Thaer's approach was facing far stronger opposition than Schulze.

Liebig's critique of the academies and the establishment of university institutes

Despite their emphasis upon practice, however, the academies had their weaknesses. Since little research was conducted there, during the 1850s most larger German states chose to establish agricultural experiment stations.[9] But the most vigorous critique of the academies was delivered in 1861 by Justus Liebig. Thaer's conception of integrating science and practice was a positive one, Liebig conceded, but the academies had failed to realise it. Unlike Thaer himself, the academies' directors had usually been successful farmers with little understanding of the sciences. Given the academies' 'monastery-like' isolation, as Liebig put it, their staff were unable to keep up with de-

7 Becker and Schmitt, 'Entwicklung der Agrarwissenschaften'.

8 Boehm, 'Geschichte des Landwirtschaftsstudiums', 7–8; Schwabe, 'Zur Idee', 36.

9 Finlay, *Science, Practice & Politics*.

velopments in the natural sciences (not least of which was Liebig's own theory of plant nutrition and his proposal of mineral fertilisers). As a result, Liebig dismissed the academies' contribution to agricultural improvement as negligible. Agricultural education badly needed up-to-date 'science', he argued, and the only way to secure this was to integrate the academies with universities.[10]

Since various groups were sceptical of Liebig's charges, there followed a lively debate. Prussia's Agricultural Advisory Council (*Landes-Oekonomie-Kollegium*), for example, complained that Liebig knew nothing about the academies in Prussia and expressed concern that the practical orientation of agricultural education would suffer in a university setting.[11] Prussian Saxony's Agricultural Association (*landwirtschaftlicher Zentralverein*), however, had been lobbying for such an institute since 1860, and in 1862 the Prussian Ministries of Education and Agriculture agreed to establish a chair of agriculture at the University of Halle. Although Halle's rapid success may have prompted later observers to treat it as a watershed in the history of agricultural education, the transition was both more gradual and complex. For what was novel about Halle was not the chair itself; as we have seen, chairs of agriculture had existed at various universities since the early 19th century, often in faculties of 'administrative sciences' (*Staatswissenschaften*), probably as a continuation of the 18th century cameralist tradition.[12] What instead made Halle distinctive was that its chair came attached to an *institute* and that agriculture was to be offerred as a *degree* subject.

Halle's degree immediately proved attractive to students,[13] and the new reform spread rapidly. During the 1870s and 1880s similar

10 Liebig, 'Wissenschaft und Landwirtschaft', I and II.

11 Finlay, 'Death of the academies'; Giessmann, 'Einfluss der Entwicklung'; Dittrich, 'Julius Kuehn'.

12 Apart from Jena, Munich and Wuerzburg also had chairs of agriculture until the 1870s (Raum, 'Die landwirt. Hochschullehre', 473–475), and Tuebingen's chair (est. 1817) was only discontinued in 1903 (Boelcke, 'Gruendungsgeschichte', 96, 102). Berlin's chair was established in 1859 but fared poorly and was soon replaced by the College there (see Chapter Four).

13 Two years after opening, the degree programme had over one hundred students and by 1872, over two hundred (von der Goltz, *Geschichte*, vol 2, 323).

institutes were established at Leipzig (1869), Giessen (1871), Koenigsberg (1872), Goettingen (1872), Kiel (1876) and Breslau (1881), as well as at the engineering college in Munich (1874). At the same time, by 1870 nearly all of the academies had either been integrated with universities or closed. Only three remained independent – those at Hohenheim and Weihenstephan, along with a college in Vienna (*Hochschule fuer Bodenkultur*) – but they, too, were reformed so as to devote more time to science in the curriculum.[14]

Although most of the new university institutes were unable to match Halle's student numbers, all of them allocated science a large role in the curriculum. Where Thaer had regarded agriculture as a commercial subject, at the universities it was treated as an applied science. To be sure, from 1877 students in Prussia were recommended to have two years' practical experience of agriculture prior to beginning their degrees, but it was not a formal entry requirement and seems to have often been ignored.[15] Moreover only one university institute possessed a model farm (Goettingen), and although most had experimental fields, they were increasingly used for arms-length demonstrations rather than for hands-on practical instruction.[16] In addition the relation between basic sciences and agricultural subjects in the curriculum changed. Whereas at the academies first-year students encountered both sciences and agricultural disciplines in their first semester, at most of the universities students were expected to complete their science courses before moving on to agricultural ones. And as the natural sciences came to dominate the timetable, practically oriented subjects had to give way. Farm management (*landwirtschaftliche Betriebslehre*) was a case in point. Although it had

14 Boehm, 'Geschichte des Landwirtschaftsstudiums'. On the reforms at Hohenheim see Franz, *Liebig und Hohenheim*.

15 Boehm, 'Geschichte des Landwirtschaftsstudiums', 11.

16 *Landwirtschaftliche Jahrbuecher*, 1910; *Mentzel & Lengerkes Schreibkalender*, 1902. The well-known professor of animal-breeding at the College in Berlin, Hermann Settegast, was an admirer of Thaer, had criticised the closure of one of the Prussian academies and voiced his concern that most university institutes had no model farm (Settegast, *Erlebtes*).

been central to the educational ideals of the academies, it played a declining role at the universities in the later 19th century.[17]

The most prominent spokesman for this new science-dominated model of education was Julius Kuehn (1825–1910), the first professor of agriculture at Halle.[18] Having been trained in botany, Kuehn was a great admirer of science, and his major aim at Halle was to get systematic knowledge about agriculture accepted as a *science*. Thus he defined the new discipline as the 'physiology and biology of domesticated organisms'.[19] Under the circumstances then prevailing at German universities, however, gaining recognition for this new 'science' was no easy task. For one thing, the domination of the universities by the ideology of pure science and all-round learning (*Wissenschaft*) meant that the teaching of utilitarian subjects had been largely relegated to lower-status institutions, such as polytechnic schools, schools of commerce, and agricultural academies. Moreover, at Halle and the other universities, the new institutes of agriculture were established, not in faculties of science (of which there were very few before the First World War), but in faculties of 'philosophy' which embraced both the sciences and the humanities. And by and large, in humanities disciplines the ideology of *Wissenschaft* was especially strong. As a result, agriculture at more than one university was not taken seriously as a science right up to the First World War.[20]

Predictably, therefore, Kuehn ran into considerable opposition at Halle during the early years (one colleague was accustomed to call him the 'professor of manure' behind his back). Under the circumstances, one can better understand why Kuehn urged his students to attend humanities courses in order to acquire 'all-round learning'; or

17 Schwabe, 'Zur Idee', 51–55. For Gustav Drechsler (professor of agriculture at Goettingen), for example, chemistry was the foundation of the curriculum (Becker and Schmitt, 'Entwicklung der Agrarwissenschaften').

18 Schwabe, 'Zur Idee'; Boehm, *Biographisches Handbuch*.

19 According to his successor at Halle, Kuehn had always been keen on botany and would have been just as happy to be called to a chair of botany in 1862 (Wohltmann, 'Das landwirtschaftliche Institut', iv).

20 For Halle see Conrad, *Lebenserinnerungen*, 153; on Koenigsberg see Mitscherlich, *Lebenserinnerungen*, 18; on Breslau see von Ruemker, 'Landeskulturhochschulen?'.

why he required them to take the same science and economics courses as attended by other groups of students, rather than calling for tailor-made courses (eg, 'agricultural chemistry', 'agricultural botany'). And although Kuehn eventually managed to get agricultural science recognised as a dissertation subject, his success may have owed something to the fact that for many years he forbade his doctoral students to work on topics which could not be studied in the laboratory.

It would be wrong to say that Kuehn was indifferent to the practical needs of the agricultural community. His career trajectory, for example, was far more like Aereboe's than Mitscherlich's (see Chapter Two), not least because he had spent nearly twenty years working in practical agriculture before coming to Halle,[21] and some of his research at Halle was devoted to local agricultural problems, such as those of Prussian Saxony's numerous sugar-beet growers.[22] But the circumstances at Halle seem to have prompted a change of orientation. When he endorsed the 'unity of theory and practice' there, for example, he meant something quite different than had Thaer. Rather than regarding science and practice as equal and *complementary* partners, both of which constituted 'agricultural science', Kuehn adhered to a hierarchical view in which practical problems could be solved simply by the *application* of basic scientific knowledge.[23] There was nothing so practical, he noted, as a good theory. In short, Kuehn regarded himself in the first instance as a scientist and only secondarily as a servant of agriculture.[24]

21 Mueller and Klemm, *Im Dienste der Ceres*, 208–209.

22 Finlay, 'Theory versus practice'.

23 The enormous confidence in the practical power of basic knowledge has been cited as one of the reasons for the slow development of farm management at this time. For German economists (*Nationaloekonomen*) believed that there was no need for farm management as an independent discipline since the managerial problems of the farm would be solved by applying the principles of economics (Achilles, *Deutsche Agrargeschichte*, 369).

24 Though one of Kuehn's colleagues at Halle emphasised the extent to which Kuehn addressed both economic and practical considerations in his teaching, he also acknowledged that Kuehn's experiments in animal-breeding and plant pathology were directed primarily at understanding basic processes rather than solving practical problems (Bode, 'Vom Geist Julius Kuehns', esp 123).

Calls for reform from the 1890s

The late 19th century was a turbulent period in the history of German higher education. That the classic early 19th century Humboldtian ideal of the university – as an autonomous community of scholars engaged in 'pure' science – was beginning to come under pressure can be seen in the intense debates over the proper relationship between technical education and the universities.[25] From the 1870s, for example, supporters of the engineering colleges began to demand parity of status with the universities (which the latter vigorously opposed). This was also a period when universities came under fire for neglecting disciplines of industrial significance, though they continued to resist the incorporation of the 'applied sciences' with remarkable tenacity. One consequence of this resistance was that local groups of businessmen in some cities began to join with municipal authorities, not only to establish independent commercial colleges but eventually also to set up private universities of a more 'modern' character.[26] By the 1890s, however, even the engineering colleges were beginning to feel the pressures to reform. Colleges were accused of having allowed too much advanced mathematics into the curriculum and of ignoring industrial needs. Engineering staff generally agreed on the desirability of students spending 'sandwich' periods in industry, on greater emphasis on economics and workshop-practicals in the curriculum, and that professors ought to have experience of industry or commerce. With an eye to the continuing dominance of the ideology of 'purity', however, they were hesitant to go further for fear of compromising their academic status.[27]

Discussion of the reform of agricultural education at this time covered much of the same ground. By the 1890s a number of academics had begun to voice misgivings with the Kuehnian model of agricultural education, often on the grounds that it was not sufficiently practical. Some critics complained that the humanities courses which

25 Manegold, *Universitaet, Technische Hochschule und Industrie*.

26 See the literature reviewed in Harwood, 'Institutional innovation'.

27 Eg, Gispen, *New Profession*; Gundler, *Technische Bildung*, 87ff.

Kuehn had recommended were of no use to farmers while others wanted more attention paid to agricultural courses and less to basic sciences. And many complained that farm management was not being given an adequate place in the curriculum.[28]

The most important of these critics was undoubtedly the professor of plant-production at the University of Breslau, Kurt von Ruemker (1859–1940). Well-acquainted with Kuehn's views, having received his doctorate with him, von Ruemker began in the mid-1890s to publish a series of proposals for the reform of agricultural education at the universities. Although, as we shall see, he agreed with Kuehn on some issues, von Ruemker was more fundamentally an admirer of Albrecht Thaer. His major criticism of the Kuehnian model was that the classic ideal of all-round learning had become unrealistic. Unlike the academies, where there were specialised professorships in most agricultural disciplines, the university departments at that time usually had only a single professor who taught all of the agricultural disciplines himself. But the rapid growth of knowledge in agricultural fields since the 1860s, von Ruemker argued, meant that trying to maintain this kind of one-man show was hopeless and could only lead to dilettantism. It was essential, therefore, that new specialised positions in the various agricultural sciences be established at the universities.[29]

At the same time he was very much aware that agriculture was not particularly welcome within the Philosophical Faculty. Thus he devoted a fair amount of ink to challenging the grounds on which 'applied' subjects were looked down upon within the universities. Were the humanities really taught purely for their own sake, he asked; wasn't a knowledge of history, for example, important for politicians? And weren't medicine (and law) just as 'applied' subjects as agri-

28 eg, Waterstradt, 'Aufgaben der Landwirtschaftswissenschaft'; von der Goltz, *Geschichte*, vol 1, 23; Seelhorst, [review of Hansen]; Feige, 'Die wissenschaftliche Stellung'. Interestingly, in the engineering colleges, too, management (*Betriebswirtschaftslehre*) did not find a place in the curriculum until quite late; the first chair devoted to the subject was established at the engineering college in Berlin in 1904 (Gundler, *Technische Bildung*, 97).

29 von Ruemker, *Landwirtschaft und Wissenschaft*; von Ruemker, 'Die Systematik'.

culture? Indeed, weren't the problems faced by the farmer and the doctor – as well as the character of the knowledge with which they tried to solve them – quite similar?[30] Perhaps sensing that such arguments might cut little ice, however, he also suggested in 1897 that removing the natural sciences from the Philosophical Faculty in order to create a new Science Faculty (as had been done for the first time at Tuebingen in 1863) was quite a sensible thing to do and that the new Faculty would provide a better home for agriculture. A decade later, however, he was calling instead for the creation of new 'Economics Faculties' where agriculture would be located alongside kindred disciplines like economics and forestry.[31]

Far from trying to ease the tensions within the Philosophical Faculty by playing down agriculture's practical side, as Kuehn had done, von Ruemker called instead for a stronger practical orientation in the curriculum. By concentrating upon the natural sciences to the neglect of farm management and economics, he wrote, the Kuehnian curriculum was 'one-sided', but it was also desirable that more university-institutes should acquire experimental farms where new knowledge could be produced under realistic conditions.[32] The usual entry-requirement for students of two years of practical experience in agriculture, he felt, was totally inadequate – three to four years were the minimum – and some staff also needed to maintain closer ties with the farming world than they had in the past.[33] As one might expect, therefore, von Ruemker's general conception of agricultural science owed rather more to Thaer than to Kuehn. Although he agreed with Kuehn that their subject was an applied science, von Ruemker insisted

30 von Ruemker, *Landwirtschaft und Wissenschaft*, 16–17, 28–30; von Ruemker, 'Gehoert die Landwirtschaftslehre?', 214, 216.

31 von Ruemker, 'Was konnen wir aus der Geschichte?', 727–728; von Ruemker, 'Zur Frage der Ausgestaltung'.

32 On the neglect of farm management see von Ruemker, *Landwirtschaft und Wissenschaft*, 19–20; and his 'Die moderne Landwirtschaftswissenschaft'. On experimental farms see his 'Ueber die Bedeutung', 194ff.

33 On students' work experience: von Ruemker, *Ausbildung des praktischen Landwirts*. On staff's need for closer links to practice see von Ruemker, *Landwirtschaft und Wissenschaft*, 20–21; his 'Ueber die Bedeutung', 192; and his 'Das landwirt. Versuchs- und Bildungswesen', 104.

that it drew equally upon both the natural sciences *and* economics. For the point of agricultural science, he emphasised, was to help the farmer to achieve his economic aim: to maximise the return on investment. The foundation subjects in the curriculum were, therefore, farm management (drawing upon economics), plant production, and animal production (both of which drew upon the natural sciences).[34]

Though von Ruemker was critical of the universities, however, his aim was not to champion the academies. On the contrary, he had some fairly harsh things to say about the latter which are revealing, about both the structure of the 'academic field' and the wider social context in which it was situated. In a paper aimed explicitly at members of the current and future agricultural elite (owners and directors of large estates and agricultural administrators) von Ruemker placed great emphasis upon the status-value of a university education.[35] Small farmers, he argued, could get away with an exclusively practical training, but a well-rounded education was essential for the elite, and only the universities were equipped to offer this.[36] To begin with, in order to acquire a fundamental understanding of the relevant sciences,

34 von Ruemker, 'Was konnen wir aus der Geschichte?'

35 In doing so he was up against a fairly sceptical group of customers. For relatively few farmers sent their sons for higher agricultural education. During the 1920s among the sons of *small* farmers, only about half (of those educationally qualified) attended an agricultural secondary school of some kind while less then one percent went on to higher education (Klemm, *Agrarwissenschaften in Deutschland*, 285; Titze, *Datenhandbuch*, Table 127). Among the owners of large estates in the late 19th century only about 20% saw fit to send their sons for higher education (Settegast, *Erlebtes*, 205). By 1920 the proportion had increased to about half, but of those sons who did turn up on campus, the great majority left after a few semesters without completing a degree (Oldenburg, *Der Ausbau*, 17). The situation in the United States was similar; during the 1920s only about one percent of U.S. farmers' sons went to college (Shepardson, *Agricultural Education*, 60).

36 He had to concede that not all of the academies were deficient in this respect. Students at Bonn and Berlin, for example, were not so narrow since they had access to courses at the associated universities (von Ruemker, 'Die Ausbildung des praktischen Landwirts'). Von Ruemker was by no means the only person to stress the superiority of a well-rounded university education; at Munich much the same arguments were then being made to justify relocating agricultural education to the University (Wollny and Pott, 'Antrag').

the agricultural student – just like the medical student to whose status he aspired – needed to take *general* science courses rather than ones tailored to agricultural problems. To look for an immediate payoff from science courses, he wrote, was 'unworthy' of a university student; during their lives such individuals would always lag behind those who had a proper education.[37] Furthermore students aiming to join the agricultural elite would benefit substantially from hearing lectures in the humanities since these 'elevated' the mind and fostered a broad perspective. Referring perhaps to estate-owners' unease at the rise of a new and powerful bourgeoisie in Germany, von Ruemker closed this paper by saying that if the agricultural elite were to stand fast in the 'struggle for existence with other educated classes', a general education was crucial. Knowledge was power, he insisted, and it was perhaps the only effective counterweight to the power of money.[38]

Despite the attention which von Ruemker drew to the university-institutes' inadequacies, therefore, his reform-proposals were evidently designed to strengthen them while marginalising the academies. If some of the academies' more specialised and practical features could be appropriated without abandoning the universities' stronger 'scientific' character, the universities could have their cake and eat it. They

37 von Ruemker, *Die Ausbildung des praktischen Landwirts*, 23.

38 eg, von Ruemker, *Die Ausbildung des praktischen Landwirts*, 26; von Ruemker, 'Gehoert die Landwirtschaftslehre?', 214. This kind of direct appeal to the class interests of estate-owners was also common after the First World War (see section four). Although von Ruemker spoke mainly of the cognitive advantages of a university education, his listeners were probably just as attracted by its social prospects. Certainly those German businessmen who championed the establishment of commercial colleges had social advantages in mind (Franz, *Zwischen Markt und Profession*, chp. 2). And Terry Shinn has argued that one of the reasons why middle-class French families in industry or commerce preferred to send their sons to the Ecole Centrale rather than to a more practically oriented ‚lower' ecole was not just that its education was potentially useful but that it would increase their social prestige and help them to make contacts (Shinn, 'From corps to profession').

would then be just as 'practical' as the academies while conferring greater prestige.[39]

Perhaps because von Ruemker's proposals constituted quite a radical agenda for reform, few of them were adopted until the interwar period. The issues which he raised continued to be discussed in the agricultural journals before 1914 – there were occasional criticisms of the Kuehnian model or the academies, as well as attacks upon von Ruemker's own proposals – but no sustained discussion, and he was disappointed that Kuehn himself hadn't bothered to reply.[40] Just why these proposals fell on deaf ears is not clear. One of the few academics to voice support for von Ruemker suggested that the need for greater specialisation among agricultural staff was widely recognised within the universities but that professors had declined to draw attention to it 'on opportunistic grounds' (most likely a reference to the fact that professors' income from capitation fees would have been hurt by the creation of new chairs).[41] But it probably also had something to do with the fact that the period from the late 1890s to 1914 is generally regarded as a 'golden age' for German agriculture. Tariffs protected farmers from foreign competition while industrialisation spurred demand for their produce. Furthermore with farming in good shape,

39 In a remarkable volte-face in 1909, however, he abandoned his support for (suitably reformed) universities as the natural home for agricultural education, probably because of the frustration he faced in introducing changes at Breslau (see Chapter Four). The only viable institutional locations, he now argued, were large independent colleges – equivalent in size and status to universities and engineering colleges – which contained agriculture, veterinary medicine and forestry (von Ruemker, 'Ueber die Entwicklung' and 'Landeskulturhochschulen').

40 For a critique of the Kuehnian model by a defender of the academies, see Waterstradt, 'Aufgaben der Landwirtschaftswissenschaft'. For a defense of the university see Feige, 'Die gegenwaertige Gestaltung' and Wohltmann, 'Die Festfeier'. For a tribute to Kuehn see Kirchner, 'Landwirtschaft und Universitaet'. On von Ruemker's critics see his *Landwirtschaft und Wissenschaft*, preface. On Kuehn's failure to reply see von Ruemker, 'Die Systematik', 409.

41 Schindler, 'Ueber Ziele', 237–238; Boehm, 'Geschichte des Landwirtschaftsstudiums', 13.

student numbers were buoyant.[42] Under the circumstances there was little reason for either farmers or academics to call for wholesale educational reform.

Nevertheless, with expenditures on higher education increasing in all of the German states, the climate was favourable for a limited amount of innovation. Some of these changes occurred within the university sector. Although Friedrich Althoff, the senior civil servant in charge of higher education at the Prussian Education Ministry, was appalled at the prospective cost of implementing von Ruemker's recommendations throughout Prussia, he was at least prepared to help out von Ruemker's own university (see Chapter Four). Thus by 1899 the University of Breslau had acquired separate institutes for agronomy, animal husbandry, agricultural bacteriology and agricultural engineering (other universities had to wait until the 1920s). Growth and differentiation also occurred outside the university sector. The Agricultural College at Berlin, for example, became much larger and more specialised than any other academy or university institute.[43]

Over the two centuries of its existence, therefore, German agricultural education had undergone a series of experiments. Designed according to widely differing visions of what 'agricultural science' should be, the curriculum veered alternately between the 'scientific' and the practical pole. That is not to say, of course, that such fluctuation was simply cyclical. Even had they wanted to, educators were not free to retrace their steps, in part due to the demands for innovation coming from farmers, ministries, and elsewhere, but also because the body of knowledge to be conveyed was constantly growing. Moreover, as we have seen, by the First World War the higher education system had diversified, embracing institutions which varied greatly in size, structure and orientation toward science/practice. Given the favourable climate of the time, such diversity was un-

42 Between 1880 and 1914 student numbers increased markedly at all institutions, from two-fold at Halle to more than ten-fold at Berlin (Hansen, *Das landwirtschaftliche Unterrichtswesen* (1920), 19).

43 By 1906 the Berlin Agricultural College had 18 professorial staff (ie, tenured full and associate professors) and 19 separate institutes vs. Halle's 7 professorial staff (Klemm, *Agrarwissenschaften in Deutschland*, 215).

problematic and had the merit of offering choice to students. All of this, however, was soon to change.

Controversy in the aftermath of war: academies versus universities, 1919–1920

Germany's defeat in the First World War brought severe economic crisis and political instability in its wake. The new social democratic government naturally brought with it a set of priorities different from the former monarchy's, but the traumas of the preceding four years also forced a rethink of longstanding policies. By 1918 German agriculture was in a desperate state. Animal stocks were depleted; soils were impoverished due to the wartime diversion of nitrates from fertiliser-manufacture to explosives; agricultural regions had been lost via the Versailles Treaty; and machinery was in bad shape for lack of maintenance and investment.[44] Moreover the loss of food imports during the war had made for serious malnutrition among substantial sections of the German populace,[45] and it was unclear whether the country would be able to rely on foreign suppliers after the war. As a result there was widespread agreement that agricultural productivity needed to be increased substantially and quickly so as to restore some kind of normality in the near future and to secure Germany's economic independence over the longer term. This could only be achieved through the adoption of intensive farming methods. Not surprisingly, therefore, the organisation and expansion of agricultural education was high on the government's agenda.[46]

44 Heim, *Kalorien*, introduction; Klemm, *Agrarwissenschaften im Dritten Reich*, 7–14. In 1918–19, for example, the consumption of nitrate- and phosphate-fertilisers in Germany was roughly half what it had been on the eve of war (Welte, 'Bedeutung der mineralischen Duengung', 48).

45 Offer, *First World War*.

46 The argument for independence from food imports was already being advanced within the Prussian Agriculture Ministry before the war (Min. Landwirtschaft to

In Prussia the education debate in governmental circles was triggered early in 1919 by what seemed like a modest request: the Agricultural Ministry asked the Education Ministry to grant the academy at Bonn the right to award the doctorate. This would have put Bonn on a par with the Agricultural Colleges at Berlin and Hohenheim which had been granted this privilege the previous year. The Education Minister (under whose jurisdiction this issue fell) agreed but with the proviso that the Bonn Academy be incorporated into the University of Bonn as an 'agricultural faculty'.[47] This proviso was in keeping with the Prussian Ministry's overall policy for the reform of higher education after the war, one of whose aims was to break down the status hierarchy within higher education. In 1919 C.H. Becker (who became Prussian Minister of Education a few years later) declared that the segregation of pure and applied scholarship, which since the latter 19th century had placed engineering and commerce in lower-status non-university institutions, had been a mistake.[48] His call to integrate the engineering colleges with the universities got nowhere, though a less radical proposal found considerable support within the engineering community: namely, to integrate colleges of commerce and of agriculture into the engineering colleges in order to create unified 'colleges of technology and commerce' (*Hochschulen fuer Technik und Wirtschaft*).[49]

But the Agriculture Minister was not enthusiastic. Faced with the need to boost agricultural productivity, he insisted that Prussia's two agricultural colleges (at Bonn and Berlin) had to remain independent institutions under his jurisdiction on the grounds that at a time of national economic emergency such as then prevailed, agricultural education needed a strong tie to 'practice' which could not be maintained in a university. Indeed, the logic of this situation, he argued, meant that even the university institutes of agriculture ought to be under the

Min. Wissenschaft, 11 April 1914, Rep 77, Tit. 46, Nr. 31, GStaA; see also Oldenburg, *Der Ausbau*) as well as by many academics after 1918 (see below).

47 Min. Wissenschaft to Min. Landwirtschaft, 27 Feb 1919, Rep 77, Tit. 46, Nr. 31, GStaA.

48 Becker, *Gedanken zur Hochschulreform*.

49 Duewell, 'Die Neugestaltung'.

Agriculture Ministry rather than that for Education.[50] For his part, the Education Minister seems to have agreed that the Philosophical Faculty was not a satisfactory location for university-institutes, but he was unwilling to relinquish his control over teaching and appointments at them. Evidently hoping to win over the Agriculture Minister to his plan, he suggested various power-sharing arrangements in which both Ministries would oversee the new 'Faculties of Agriculture'.[51]

While ministers exchanged views in private, a public discussion was also taking place over the need to restructure agricultural education. And the public debate brings out very clearly the contrasting visions of agricultural education which had been institutionalised before the First World War. At the centre of discussion was a book by the professor of agriculture at the University of Koenigsberg, Johannes Hansen.[52] If the productivity of German agriculture were to be increased, Hansen argued, German farmers would need to be better educated. To that end he proposed a major expansion of specialised university institutes of agriculture, thus echoing one of von Ruemker's earlier proposals. But unlike the Prussian Education Ministry or von Ruemker, Hansen had no reservations about the value of the academies. Liebig's critique, he wrote, had been exaggerated and ill-informed, and the academies' possession of experimental farms was a decided advantage. Moreover, in Prussia the academies were in general better equipped with teaching and research facilities than the universities, and the universities, he felt, ought to tailor basic science courses to the needs of agricultural students, as was customary at the academies.[53]

Given the high political profile of agricultural matters at that time, Hansen's book found a ready audience and was sold out within

50 Min. Landwirtschaft to Min. Wissenschaft, 11 April 1919 and 16 June 1919, Rep 77, Tit. 46, Nr. 31, GStaA.

51 Min. Wissenschaft to Min. Landwirtschaft, 18 July 1919, Rep 77, Tit. 46, Nr. 31, GStaA.

52 Hansen, *Das landwirtschaftliche Unterrichtswesen* (1919).

53 Significantly, proposals similar to Hansen's were then being made for the reform of engineering education in order to bring it closer to 'practice' (Duewell, 'Die Neugestaltung').

eight months and reprinted within the year.[54] His call for university reform, combined with his favourable treatment of the academies, seems to have met with agreement, not only from academy professors but even from a few at the universities.[55] Erwin Baur, the outspoken professor of genetics and plant-breeding at the Agricultural College of Berlin, was typically blunt in his analysis of the problems. The Prussian Agriculture Ministry, he complained, had totally failed to devise a rational scheme for agricultural education and research. Had there been a sensible division of labour among existing institutions, current funding (albeit miserly) would have been far more effective, and there would be no need for new institutions.[56] Despite calling for rationalisation, however, Baur said nothing about the relative merits of academies versus universities as educational institutions. As late as the spring of 1919, therefore, there was no talk of closures or amalgamation, and neither type of institution seems to have felt under threat. Indeed, in the first edition of his book (completed in December 1918 and published in the spring of 1919) Hansen declared that the 'old dispute' between academies and universities was now closed.

By the autumn of that year, however, it was clear that Hansen had badly misjudged the situation. Details of the ministerial discussion had by then leaked out, and rumours were circulating that the Prussian Agriculture Ministry wanted to close some university institutes and to remove others from the university and bring them under its control so that new agricultural colleges could be established. Although I have found no ministerial documents to confirm these rumours, by mid-November the agricultural press was no longer referring to 'alleged plans' but to specific proposals which the Agriculture Minister had submitted to the cabinet.[57] Moreover, fears within the universities that the Agriculture Ministry favoured the academies were probably augmented by the fact that the Minister was at that very moment upgrading the Bonn Academy into a college and granting it the right to

54 Hansen, *Das landwirt. Unterrichtswesen*, (1920), preface.

55 Seelhorst, 'Das landwirt. Unterrichtswesen'; Kraemer, 'Das landwirt. Unterrichtswesen'; Remy, 'Landwirtschaftliche Hochschulen?'.

56 Baur, 'Die Zukunft der preussischen'.

57 Hansen, 'Landwirtschaftliches Universitaets-Institut'.

award the doctorate.[58] In any event, for the next eight months the agricultural press (and to some extent the daily press) were peppered with articles by academics, estate-owners and students, arguing the merits of academy versus university. Lurking behind virtually all of the contributions to this debate was the classic question once again of science versus practice: was the primary aim of such education to increase agricultural productivity or to confer a fundamental understanding of agricultural processes?[59]

A few contributors to the debate (like Hansen before them) adopted a mediating stance, arguing that both academies and university-institutes had a role to play and that the key issue was to fund agricultural education more generously in order to improve both kinds of institution.[60] But for the most part the debate was polarised, waged by academics defending either university or academy.[61]

Central to the defence of the universities was the importance assigned to a well-rounded education (*Allgemeinbildung*). This, university professors believed, could only be obtained at a university. Moreover incorporating the agricultural colleges into the universities

58 Remy, 'Das hoehere landwirt. Unterrichtswesen'.

59 As I pointed out in Chapter One, these issues are common to all of the 'practical sciences'. Around the turn of the century, for example, the proponents of higher commercial education in Germany were debating whether this kind of education was better conducted in specialised university faculties or in independent commercial colleges (Franz, *Zwischen Markt und Profession*, 36–37). And in France between 1895 and 1914 there was a similar clash over the organisation of engineering education in which staff at the newer industrially-oriented grandes ecoles des sciences appliques called for the closure of the older, more prestigious and heavily mathematically oriented Ecole Polytechnique (Shinn, 'From corps to profession', 204–206).

60 Gerland, 'Nochmals'; Martiny, 'Zur Frage des hoeheren'; Hansen, 'Landwirtschaftliches Universitaets-Institut'. Predictably this was also the stance adopted by the association representing German agricultural students, declaring that university departments and colleges complemented one another, both having 'proved themselves' (Gerken, 'Die Tagung').

61 The small number of non-academics who expressed a view on this issue seem to have been divided over the matter (Schroeder, 'Hochschule oder Universitaetsinstitut?'; Elleke, 'Noch einmal'; Kluge, 'Das Hochschulstudium').

would be far cheaper than creating new colleges.[62] Along similar lines, estate-owners who had studied at universities, along with the Association of Agricultural [school] Teachers, protested any closure of university institutes, insisting that a general education was essential to their status and authority as spokesmen for agriculture.[63] Some professors argued that removing agriculture from the universities would harm agricultural research, both by eliminating the cross-fertilisation with other disciplines at the university and by subjecting it to the destructive controls of the Agriculture Ministry.[64]

As might have been expected, the academy professors took a different line, generally stressing the academies' superior facilities. One noted that the cost per student of an agricultural education was lower at the academies or colleges than at the universities.[65] More surprising perhaps is that quite a few also accepted the premise that general education was important, though they insisted that this could be obtained just as well at an agricultural college, providing that such colleges were in the vicinity of a university town. These particular academy professors were thus defending the merits of 'well-placed' colleges (ie, at Berlin and Bonn) while tacitly condemning the 'isolated' ones (at Hohenheim and Weihenstephan).[66]

But a few academy professors adopted a more radical and utilitarian stance, challenging the educational premises which under-

62 Mitscherlich, 'Das Studium der Landwirtschaft'; Wygodzinski, 'Universitaeten'; Froelich, 'Nochmals'.

63 Heumann and von Campe, 'Landwirtschaftliche Fachhochschulen'; Woelfer and Lentz, 'Landwirtschaftliche Hochschulen'; Elleke, 'Noch einmal'; Boehmer, 'Zur Frage'; Hoppe, 'Die Ausbildung'. Hansen argued along the same lines in his *Das landwirt. Unterrichtswesen* (1920), 116–117. Others voicing this view, like Dr Paul Graebke, seem to have been neither farmers nor academics but recited all the classic defences of *Bildung* and the critique of crass materialism and specialisation (Graebke, 'Die Frage'). University staff brusquely rejected Hansen's view that the traditional *Abitur* was less important than additional practical experience in preparing agricultural teachers for higher education (Woelfer, 'Nochmals'; Kerschowsky, 'Das landwirt. Unterrichtswesen').

64 Abderhalden, 'Ein Vorstoss'.

65 Remy, 'Das hoehere landwirt. Unterrichtswesen'.

66 Remy, 'Landwirtschaftliche Hochschulen?'; Lemmermann, 'Das Studium'; Martiny, 'Zur Frage der Beseitigung'.

lay the university's position. For example, Friedrich Aereboe (then professor of farm management at Hohenheim) argued that the much vaunted Kuehnian model was partially responsible for the lack of economically literate farmers during the First World War. Basic science instruction at the universities was too far removed from agricultural issues, and the location of agricultural institutes in Philosophical Faculties was a definite hindrance in developing close ties to practice. For all the talk about the importance of general education, Aereboe insisted, very few agriculture students, even at the universities, actually attended such lectures. General education could never take the place of a good technical education (*Fachbildung*).[67] In a similar vein, Theodor Remy, professor of agronomy at the college in Bonn, reacted impatiently to the argument that agricultural colleges would not recruit good staff or the best qualified students since the colleges' status was inferior to that of the universities. It was not the state's task to confer status upon agricultural students and staff, he argued, but rather to advance agriculture. Nor was it justified to cast aspersions upon the educational qualifications of agriculture students; what they lacked in formal schooling was compensated for by their greater experience of real life.[68] Hans Zoerner, a student at the academy in Hohenheim, was also sceptical of 'general cultivation [which] has become the university-institutes' battle-cry at the moment'. What, he asked, ought we to regard as the criterion of 'so-called general cultivation'? From the humanistic standpoint a person who knew little about Cicero or Horace was regarded as uncultivated, but those who counted as 'cultivated' were not expected to know anything about, say, Robert Koch or about the simplest processes of everyday life. And why were the classics thought to confer 'general cultivation' but not subjects like economics or agricultural politics?[69]

Such a challenge from the academies provoked, predictably enough, a defence of the traditional concept of 'cultivation' by staff and students at the University of Halle. Gustav Froelich, professor of

67 Aereboe, 'Landwirt. Hochschulen oder?'.

68 Remy, 'Das hoehere landwirt. Unterrichtswesen'.

69 Zoerner, 'Reform'.

animal breeding there, was quick to respond.[70] Aereboe's complaint, that basic science instruction was too general to be of use to agriculture students, Froelich countered, was 'one-sided', and the claim that Philosophical Faculties exercised an undesireable influence upon appointments was out of date. In response to Aereboe's rejection of the value of 'general cultivation', Froelich seized the cultural high-ground, declaring that 'no objective discussion is possible with those who lack any understanding or sense of the intangible advantages' of a broad education. Agricultural students at Halle similarly sought to distance themselves from the utilitarian recommendations for curricular reform which had been put forward by the students' association at Hohenheim and endorsed by the national association.[71] Halle's students repeated all the familiar arguments in favour of general education. They did not want a reduction in the amount of time spent upon basic sciences; did not want basic sciences to be taught from an agricultural point of view; and insisted that the *Abitur* was an essential entrance requirement for anyone wishing to take the six-semester degree. One must distinguish clearly, they emphasised, between the practical educational needs of those wishing to work as farmers and the needs of other students who sought a scientific training in preparation for careers as teachers, organisers, advisors or scholars. The utilitarian recommendations emanating from Hohenheim, they complained, had damaged the reputation of agricultural students in the eyes of other university students at a time when the former were fighting for equality of status. In reply, Hohenheim's students belittled their counterparts' preoccupations with social esteem.[72] Obviously the student body was anything but united in their conception of what an agricultural education was for.

After the vigorous public exchanges of 1919 and 1920, the controversy soon subsided (though the tensions between ex-academies

70 Froelich, 'Nochmals'.

71 Biederbeck, 'Die Tagung', objecting to the portrayal of students' views in Flaig, 'Eine ausserordentliche'.

72 Flaig and Zoerner, 'Nochmals'.

and universities continued to rumble on during the 1920s[73]). And none of the radical changes which had been considered in the immediate postwar years were implemented during the Weimar Republic (except in Bavaria; see Chapter Five). What happened instead was incremental change. Over the next decade increased expenditure on the university-institutes, along with organisational changes in the academies, brought about not only an expansion of the higher education system as a whole but also a growing resemblance between universities and academies.

Expansion, convergence and crisis during the 1920s

Although post-war financial stringency ruled out new initiatives in most sectors of German higher education, agriculture was a major exception. For one thing, student demand for agricultural education was buoyant; in the winter of 1922–23 there were more than twice as many students of agriculture at German universities than in the winter of 1913–14.[74] This, along with the need to boost food production, meant that after 1918 ministers were prepared to play a more interventionist role in agricultural education (and were perhaps also under more pressure to do so in the democratic climate of the new Weimar Republic). Initially, as we saw above, they explored forms of radical reorganisation without implementing them. Thereafter, more resources were poured into the system, pushing some institutions in a more practical direction while upgrading others. The result, whether intend-

73 In his memoirs Friedrich Aereboe claims that he agreed around 1924 to write the entry on agricultural education for a major reference work so that others would *not* be invited to do so (Aereboe, *Jugendlust*). Thereafter he seems to have taken more than one opportunity to defend the colleges against their university-critics (eg, Zoerner and Aereboe, 'Landwirtschaftliches Unterrichtswesen'; Aereboe, *Agrarpolitik*; Aereboe, 'Landwirtschaftliche Hochschulen', 641). But Aereboe was not the only one who continued to have strong feelings on the matter (Schroeder, *Das Studium der Landwirtschaft*, 4; Ausschuss zur Untersuchung, 'Das landwirt. Bildungs- und Beratungwesen').

74 Table 65, Titze, *Datenhandbuch*.

ed or not, was to erode much of the institutional diversity which had characterised the pre-1914 period.

At some places expansion simply enlarged and improved the quality of existing facilities. Breslau, for example, acquired a well-equipped new building to house all of its many institutes (the Institute for Plant-Production alone getting 18 rooms) while Koenigsberg's Institute of Agronomy got a new building with excellent laboratory facilities. Elsewhere universities acquired more extensive demonstration facilities. Animal-breeding at Koenigsberg and Halle, for example, obtained large experimental farms while Goettingen was able to establish a new institute for farm management and took over both an experiment station and an experimental farm from the local chamber of agriculture. Though attempts to establish agriculture at a wider range of universities failed, one of von Ruemker's reform proposals from the 1890s was finally implemented. Goettingen, for example, adopted the Breslau-model of multiple institutes/chairs and collegial chairmanship, dividing its existing institute of agriculture into three specialised ones, and the same occurred at Halle, creating five new institutes out of the former one. By 1930 the traditional one-man institutes of agriculture had been replaced at nearly all of the universities with separate chairs and institutes for each of the three fundamental disciplines: agronomy, animal husbandry, and farm management.[75]

This expansion and reorganisation had important consequences for higher agricultural education, both quantitative and qualitative. For although the volume of activity and the degree of specialisation at many institutions increased, the *character* of the teaching and research

75 On Breslau see Zorn, 'Die neuen landwirt. Institute'. On Koenigsberg see box 1, file 33, CB Hutchison papers. On Goettingen see Herpel, *Die Entwicklung*; Becker and Schmitt, 'Die Entwicklung der Agrarwissenschaften'. On Halle see (Froelich, 'Das Tierzucht-Versuchsgut', 1926) and (KultMin to FinMin 13 Aug 20, Rep 151, IC, Nr. 6829, GStaA. On improved facilities at the colleges see Chapter Four. On the failure to establish agriculture at Greifswald and Muenster, see Nr. 12507 and 12506, Rep 87B, GStaA. On the overall growth of chairs see *Mentzel und Lengerkes Schreibkalender* for 1913, 1920, and 1925 and von Ferber, *Entwicklung des Lehrkoerpers*, 199. By 1930 only Kiel and Giessen lacked one or other of the three basic chairs (Ausschuss zur Untersuchung, *Das landwirt. Bildungs- und Beratungwesen*).

at individual institutions also changed. Those institutions which had been 'scientifically' inclined before the War (mainly universities) acquired improved demonstration facilities while the heretofore practically oriented academies acquired fuller provision for basic sciences and were upgraded to university status. The result was a process of convergence in which the marked contrasts between scientific and practical institutions, which had accumulated during the pre-war decades, began to fade, thus substantially narrowing the gap between academies and universities.[76]

What did the process look like in detail? Among the science-oriented institutions which began to shift in a practical direction was the College at Berlin. Staff's pre-war complaints about the College's dearth of demonstration facilities finally led to the relocation during the 1920s of some of its institutes to the outskirts of the city where – for the first time – they acquired experimental fields (cf. Chapter Four). In addition, the appointment of a few professors with strong views on the importance of 'practice' (eg, Friedrich Aereboe in farm management, Johannes Hansen in animal breeding) suggests a post-war shift in ministerial policy toward Berlin.[77] At Munich, similarly, the newly appointed professors of agronomy and animal husbandry insisted – as their predecessors had not – that their institutes must be equipped with experimental fields (cf. Chapter Five). And at Goettingen, Breslau and Koenigsberg efforts were made to tailor the teaching of basic sciences to the needs of agriculture students.[78]

76 There are signs of a similar convergence in engineering education at about this time. Kranakis detects a growing similarity between French and American approaches to engineering after the turn of the century as neither the mathematical emphasis of the former nor the empirical emphasis of the latter were any longer regarded as sufficient (Kranakis, *Constructing a Bridge*, 295–297). The Ecole Polytechnique, for example, responded to criticism of its heavily mathematical curriculum by increasing the amount of time devoted to laboratory work and engineering subjects (Shinn, 'Reactionary technologists').

77 Attempts to fill chairs in genetics, botany and animal-breeding in the late 1920s suggest that the Prussian Agriculture Minister (and sometimes Berlin's staff) wanted appointees who were more practically oriented (Nr. 20096 and 20095, Rep 87B, GStaA).

78 Zoerner and Aereboe, 'Landwirtschaftliches Unterrichtswesen'. A meeting of agricultural officials from several states about 1920–21 is said to have agreed

Ministerial attempts to standardise higher agricultural education also hastened this trend. In 1922, for example, the relevant Prussian ministries (followed by Wuerttemberg in 1923) agreed to make two years of practical experience in agriculture a compulsory requirement for both the four-semester and six-semester degrees (rather than merely a recommendation).[79] And at the end of the decade the Prussian Agriculture Minister was advocating a number of additional measures, all of them pointing in a 'practical' direction, which would have narrowed yet further the gap between ex-academies and universities.[80]

But the shift in a practical direction was prompted not just by ministerial priorities. As often happens, rapid expansion of academe after the war made for a mismatch between supply and demand.[81] Numerous universities complained that the shortage of candidates with publications and teaching experience was forcing them to appoint agricultural scientists who had previously worked as estate managers, administrators in agricultural organisations, or teachers in agricultural schools but had little experience of academe since graduating. As a result, the new generation of academics appointed during the early 1920s probably possessed a different range of competencies – in

that henceforth basic science curricula would have to be tailored for agriculture students (remarks by Franz Matt in transcript of Landtags-debate, early 1922, Sign = I, 8, Bd.1, TUM-Archiv).

79 Benecke, *Die Pruefungen fuer Landwirte*. Until 1922 only the colleges at Hohenheim and Weihenstephan had actually required practical experience. Weihenstephan's constitution allowed it to demand one year's work experience, though several years were strongly recommended ('Schlusswort', *Bericht [...] Weihenstephan* for 1913–14). At Hohenheim the brochure for intending students stated that at least one year's experience would be necessary in order to follow the lectures in agricultural subjects (*Uebersicht [...]Hohenheim* for 1899 and 1909), though it appears not to have become a formal requirement until 1921 (minutes of meeting of the Beirat, 25 Jan 1921, AZ=3/02.12.02, UA-Hohen).

80 Min. Landwirtschaft, *Das landwirt. Bildungswesen in Preussen*, 65–86.

81 The same occurred in American agricultural research ca. 1890–1910 (Rossiter, 'The organization of the agricultural sciences').

particular a greater familiarity with the needs of practical agriculture – than did their predecessors.[82]

For their part, the academies were also changing as the long-standing process of academisation moved up a gear. One of the earliest steps in the process had been the transfer of responsibility in some states for agricultural education from the Agriculture Ministry to Education which occurred in 1865 for Hohenheim and in 1871 for Weihenstephan. A parallel step had been the upgrading of agricultural secondary schools to academies: at Hohenheim (1847), at Bonn (1880) and at Weihenstephan (1895). Subsequently the academies continued to make their way up the status-ladder. Hohenheim acquired the status of a college (*Hochschule*) in 1904, could award doctorates from 1918, and acquired a rectorial constitution in 1922. Bonn acquired all three of these features simultaneously in 1919. Last in the queue was Weihenstephan which was upgraded to a college in 1920, entitled to award doctorates from 1924, and acquired a rectorial constitution in 1925. Uniquely among these non-university institutions, Berlin had been established from the outset (1881) as a college, though even it did not acquire the right to award doctorates until 1918.[83] By the mid-1920s, therefore, there were very few *formal* organisational differences between the former academies and the universities.

The academisation process, however, also gradually shifted the ex-academies' curriculum in a scientific direction. In upgrading Weihenstephan, for example, the Bavarian Education Ministry also chose to endow it with new institutes for various basic sciences (cf. Chapter Five). And once they had acquired college status Weihenstephan and Hohenheim no longer concentrated exclusively on training future farmers via their practically oriented four-semester course

82 Hansen, *Das landwirt. Unterrichtswesen* (1919), 30; document dated 15 July 1919, file concerning appointment of Theodor Roemer, Personalakte Roemer, UA-Halle; Seelhorst, 'Das landwirtschaftliche Unterrichtswesen'. A good example was the somewhat controversial appointment at Berlin in 1920 of the agronomist, Kurt Opitz (Nr 20091, Rep 87B, GStaA).

83 Lemmermann, 'Das Studium der Landwirtschaft'; Renner, Zur Geschichte; Boehm, 'Geschichte des Landwirtschaftsstudiums'; Huber, *Deutsche Verfassungsgeschichte*, 946ff; Raum, *Beitraege*, no. 4.

but also began to offer the more science-based six-semester degree, long standard at the universities. By 1930, significantly, the first three semesters of Weihenstephan's six-semester degree were largely devoted to the basic sciences; only thereafter were students exposed to the agricultural sciences.[84] Acquiring the right to award doctorates probably prompted a similar shift of priorities since those with doctorates were necessarily trained for academe or administration rather than for farming. Despite the expansion of higher agricultural education from 1920, therefore, the range of institutional diversity within the system actually declined due to accelerating academisation. To my knowledge, no German ministry sought to preserve the pre-war diversity, perhaps by redefining academic status so that academies and other vocational institutions might have enjoyed equality with the universities *without* having to resemble them in organisational respects. 'Different but equal' was evidently not on the cards.

By the mid-20s, however, the prospects of further growth had collapsed. Although the number of agriculture students had been rising in the years prior to the First World War and even more rapidly immediately after the War, from 1924 they began to decline. By 1929 they had fallen to one-quarter of the 1924 level nationwide, and the decline continued through the early 1930s.[85] The initial decline was probably due to the deterioration of the agricultural economy from the mid-1920s which produced high levels of unemployment among estate-managers, agricultural teachers, and administrators. From 1929 this decline was aggravated by global depression.[86] As a result of the fall in student numbers, several university-institutes were threatened

84 S.Birner & H. Raum, 'Weihenstephaner Hochschulgesellschaft', 20 June 1931, Sign= Sekt II, Fach 71, Nr. 1, UA-Weih..

85 Ausschuss zur Untersuchung, *Das landwirt. Bildungs- und Beratungswesen*, 39; Titze, *Datenhandbuch*, Table 65. At Goettingen by 1932–33 student numbers were only one-tenth of the level a decade earlier (Becker, 'Von der Nahrungssicherung').

86 This was the view expressed by Weihenstephan's rector, Paul Kulisch (minutes of meeting by subcommittee, 29 Sept. 1925; Kulisch to Kultusminister, 30 Jan. 1925 and 14 May 1931, both MK 19684, BayHStA).

with closure.[87] A few mergers took place in order to rationalise costs, though in both cases agricultural or forestry colleges were incorporated into engineering colleges rather than into universities.[88] In the event, by the early 1930s no institutions had been closed.

With the National Socialists' seizure of power in 1933, however, agriculture seemed to have found a new ally. None of the university institutes previously under threat were closed; on the contrary, new agricultural institutes were established after 1939 at the Universities of Rostock and Greifswald. Staff numbers at the three agricultural colleges were markedly increased during the first year of the regime, and student numbers began to rise again. In 1936 a kind of agricultural research council (*Forschungsdienst*) was established, partly with the aim of boosting the status of the agricultural sciences. Little wonder, then, that most staff at agricultural institutions came out strongly in support of the new regime.[89] The status of these disciplines was important, on the one hand, because reducing agricultural imports (through increased domestic productivity) was crucial to the Nazi's declared aim of economic autarchy (and initially undeclared aim of waging war).[90] But it also reflected an aim among some of those in the Party who were concerned to reform higher education: namely, to bridge the social and institutional gap between 'pure' and 'applied' scholarship. Thus Party activists lobbied to get more engineers accepted for membership in the Prussian Academy of Sciences (which traditionally embraced mainly basic sciences and arts), and in the late

87 On Halle see Schwabe, 'Zur Idee', 64. On Giessen see Sign = Phil II, H1 (Landwirt. Institut, 1929–1932), UA-Giessen. On Berlin see Anon., 'Gesellschaft der Foerderer'. On Goettingen see Klemm, *Agrarwissenschaften im Dritten Reich*, 29.

88 In 1928 the college at Weihenstephan was merged with the Agricultural Faculty at the engineering college in Munich (cf. Chapter Five), and the following year the Forestry Academy at Tharandt merged with the engineering college in Dresden. In 1930 Wuerttemberg's education minister proposed that Hohenheim be merged with the engineering college at Stuttgart, but staff at both instutions protested, and in 1934 the proposal was finally dropped (Franz, *Universitaet Hohenheim*, 120–123).

89 Becker, 'Von der Nahrungssicherung'; Klemm, *Agrarwissenschaften im Dritten Reich*.

90 Heim, *Kalorien*.

1930s they established the first German academy dedicated to an engineering science (aeronautics).[91] In academe it was Nazi campaigning which forced the merger of Mannheim's college of commerce with the University of Heidelberg's institute for social sciences and of the forestry academies at Eberswalde and Hann.-Muenden into nearby universities.[92] Agriculture was similarly reorganised. In 1934 the Colleges of Agriculture and of Veterinary Medicine in Berlin were incorporated into the University of Berlin as a new Faculty. And in Bonn that year the College was reconstituted as the University's new Faculty of Agriculture.[93] With these moves the Prussian minister of agriculture surrendered his remaining authority for higher agricultural education to the ministry of education. During the Weimar period social democratic ministers had sought – in vain – to reduce the social distance between vocational colleges and universities. But after 1933 it was the Nazis – deploying the apparatus of a police-state – who made greater headway. By 1934 the only agricultural college to have survived intact was Hohenheim.

Conclusion

In this chapter we have seen how higher agricultural education originated in the 18th century university as a strand within the lecture-based and rather bookish teaching of administrative sciences. By the early 19th century agriculture was being taught as a subject in its own right, increasingly outside the university, but it was now conceived as an economic or commercial discipline, albeit one rooted in the natural sciences. Toward the end of the century, as the sciences grew in importance within the universities and began to make an impact upon the economy, the view that engineering, medical and agricultural

91 Ciesla, 'Abschied'; Trischler, 'Self-mobilization or resistance?', 77.

92 Carmon, 'Diverse and changing fortunes'; Huettermann, 'History of forest botany'.

93 Klemm, *Agrarwissenschaften im Dritten Reich.*

knowledge were derived through the 'application' of scientific knowledge seemed increasingly plausible. Many thus thought it made sense to return agriculture to the university where it could gain maximum benefit from its proximity to high quality science. But as the agricultural disciplines moved closer to the sciences, they began to lose touch with the world of farming, and by the 1890s a debate was under way over the relative merits of two competing visions of the field: the science-based model at the universities versus the practically-oriented one at those academies which had survived. Until the First World War, therefore, the higher education system embraced, broadly speaking, two distinct though complementary forms of institution. For a brief period immediately after the War the debate intensified, and it looked as though one of these forms might soon become extinct. Although this did not occur, in response to economic pressures and political priorities, the institutional and curricular diversity of the prewar era gradually declined during the 1920s, so that by 1933 only a single academy was left. In 1967 that, too, disappeared as the College of Agriculture at Hohenheim was upgraded to a University. With that event academisation had eradicated the last remaining alternative to university-based agricultural education.

In order to present the history of agricultural education in an accessible fashion, I have focussed in this chapter upon its chronological dimension, charting the rise and fall of different conceptions of the curriculum and of 'agricultural science'. For the most part, of course, this was a history of educational *ideologies*, fashioned in the scramble for resources at particular junctures when institutional reform was on the cards: the early 19th century, the 1860s, 1890s and 1919–1920. So far I have said relatively little about the actual structure of particular institutions. In the next chapter, however, we will focus upon one phase of this history – from 1890 to the First World War – and look in detail at the organisation, teaching and research of a handful of institutions. One reason for doing so is that despite the fact that the longstanding controversy over agricultural education was invariably couched in the language of 'academies versus universities', this language is somewhat misleading. For when one looks more closely, one sees that neither group of institutions was in fact entirely homogeneous: not *all* academies were 'practically' oriented, nor were all universities 'science' oriented. Following the principle 'treasure your exceptions', the next chapter aims to account for this diversity within each camp. And in doing so, we will move a step closer to understanding why over the longer term so many institutions became ever more science-oriented.

Chapter 4: Exemplars: Science and Practice at Five Institutions

In the previous chapter I traced long-term trends in agricultural education. We saw that the emphasis in the curriculum at many institutions shifted from a practical one to a 'scientific' one and then back again between ca. 1800 and the First World War. It should be clear from Chapter Two, on the other hand, that despite these general trends, at no point along the way was the system of agricultural higher education ever uniform. A good deal of variety was to be found within the system at any one time, such that some institutions can be classified as 'scientific' in their orientation while others were 'practical'. But how did those differences emerge?

In order to get some purchase on this question, we will look comparatively at five institutions in this chapter. In each case I will survey the geographical location, formal organisation, teaching programmes, staff and student backgrounds and advisory activities at each institution as well as the major elements in their politico-economic fields. In order to get some idea of the research conducted at these places, we will also look at certain general features of the research conducted there while devoting particular attention to their work in one discipline: plant-breeding. From these portraits it will be clear that two of them were broadly 'scientific' in orientation (Berlin, Halle) while three were 'practical' (Breslau, Hohenheim and Bonn). Our task will then be to identify those features common to the 'scientific' institutions which distinguish them from the 'practical' ones and which might plausibly have played a role in shaping their development (and conversely for the practical institutions).

Constructing comparisons over the *long* term, however, is quite awkward since, as we saw in the previous chapter, many institutions changed their orientation between, say, 1880 and 1930. In order to get around this difficulty, I have chosen to focus upon a more limited period: from 1890 to the First World War. This is a particularly interesting phase since it was at this time that the debate over alternative forms of agricultural education was most intense, and invidious comparisons between institutions were constantly being made by actors. The period is also methodologically convenient since it was at this time that the differences among agricultural institutions were most pronounced, making them easier to analyse. That might seem to suggest that the institutional differences studied in this chapter were untypical of thc 19th and carly 20th centuries and thus should not form the basis of an argument. But recall that the aim of this chapter is not to describe representative institutions but to analyse extreme and unusual ones. If we want to understand why colleges and university institutes often differed in their orientation, therefore, we need to pay especial attention to unusual universities (like Breslau) and unusual colleges (like Berlin) in order to figure out why they were odd. Thus in this chapter we will look comparatively at five rather different places in order to get a preliminary sense of the agents or circumstances which were most decisive in shaping an institution's orientation.

The Agricultural College at Berlin

Agriculture had been taught in Berlin since the early 19th century when Albrecht Thaer lectured at the University. Later in 1859 an agricultural institute was established in the University's Philosophical Faculty, but its status was anomalous. For one thing, it was funded by the Prussian Ministry of Agriculture (rather than by Education), but in addition relations between the institute and the Faculty were strained since the institute had no entry-requirement for students, and the Faculty feared a decline in its academic quality.[1] The situation was eventually resolved in 1881 with the establishment of a separate College of Agriculture. From the outset the College was a privileged institution.

1 Klein, 'Die Entwicklung der Agrarwissenschaften'; Papendieck, 'Die Entwicklung'; Klemm et al., *Von der Koeniglichen Akademie*, 25.

Endowed with a large number of professorships (12), it was also given the status and constitution of a 'college', in contrast with Hohenheim, Bonn and Weihenstephan which at that time were still 'academies'. It was also well connected to the Prussian state since it was given a board of overseers (*Kuratorium*) which consisted of the top civil servants in the Ministries of Agriculture (Hugo Thiel) and Education (Friedrich Althoff). The College grew quickly. Within twenty years it had twice as many students as the next largest institution (Halle), and by 1910 it had 18 chairs, over twice as many as the next largest institutions (Halle and Breslau).[2]

When one surveys the College's impressive array of institutes at that time, what stands out is the generous provision for the natural sciences. Of the ten appointments made in 1881, for example, fully eight were in basic science disciplines (three of them in botany alone) versus only two in agricultural fields. Provision for the sciences remained decidedly superior to that for agricultural disciplines right up to the First World War. By 1906 the number of botanical chairs had been reduced to two, but there were still no chairs dedicated specifically to the core agricultural disciplines, agronomy and farm management. Indeed, farm management was taught by a specialist in cattle breeding until 1912 when it finally acquired a dedicated chair and institute; agronomy followed suit a year later.[3]

An example of this science-orientation was the creation at the College's founding of an institute for animal physiology (a rarity at that time even within German universities). As the first occupant of this chair (Nathan Zuntz) remarked, the existence of this institute testified to a belief in the practical value of 'pure scientific research'.

2 On the establishment of the College see Wittmack, *Die Koenigliche Landwirt. Hochschule*; Klein, 'Die Entwicklung der Agrarwissenschaften'; Papendieck, 'Die Entwicklung'; Klemm et al., *Von der Koeniglichen Akademie*. On its development to 1910 see *Landwirt. Jahrbuecher*, vol 39, supplem. vol. II, 1910. On the board of overseers see Buelow, 'Grundlagen'.

3 On basic sciences see Wittmack, *Die Koenigliche Landwirt. Hochschule*. On farm management and agronomy see Baumann, 'Acker- und Pflanzenbau'; Klemm et al., *Von der Koeniglichen Akademie*; and *Landwirtschaftliche Hochschule Berlin: Jahresberichte*, 1892–1921.

A quarter of a century later his successor, Carl Neuberg, echoed a similar sentiment: agriculture was 'applied biochemistry'.[4]

In one respect, the ethos of basic science, combined with the College's promotion of animal physiology, certainly paid off. Eduard Buchner (professor of chemistry, 1898–1909) was awarded a Nobel Prize in 1907 for his classic work during the late 1890s on enzymes. And Carl Neuberg's work in the Institute of Animal Physiology before the First World War later won him the directorship of the Kaiser-Wilhelm Institute for Biochemistry as well as nominations for a Nobel.[5]

Although large new laboratories were being constructed for the sciences in the years before the First World War, the agricultural disciplines still had scarcely any experimental grounds. Berlin, for example, was the only one of the college/academies without a model farm. Before the War the College's only demonstration facilities were a tiny (0.6 ha) experimental field in the city as well as several small ones (5, 3 and 0.5 ha, none of which actually belonged to the College) in the suburb of Dahlem, a half hour's train journey away.[6] The consequences for teaching were dire. Describing the Institute of Zoological Technology's (*Zootechnik*) facilities, for example, its director had to concede that as he had practically no living animals, experiments or demonstrations on animal husbandry or breeding were out of the question. All that he could offer students was 'essentially theoretical instruction', based on pictures and models of the major domestic breeds, apparatus used with animals, and a collection of different types of wool. Though he didn't say so, his institute was in effect a museum.[7] Right from the start, therefore, the Ministry had evidently

4 On Zuntz see Wittmack, *Die Koenigliche Landwirt. Hochschule*, 160; Neuberg, 'Biochemie', 160.

5 On Buchner see Schriefers, 'Eduard Buchner'; on Neuberg see Nachmansohn, *German-Jewish Pioneers*, 311–327.

6 In 1906–07 a new 5-story laboratory for physics and animal physiology was agreed, as was a new chemistry laboratory in 1909–1910 *(Landwirtschaftliche Hochschule Berlin: Jahresberichte*). On experimental fields see von Ruemker, 'Derzeitige Stand des landwirt. Unterrichtswesens'.

7 Wittmack, *Die Koenigliche Landwirt. Hochschule*, 75–87. It was not until the mid-1920s that this Institute finally obtained its own experimental grounds.

decided that Berlin would have to measure up against the university departments (with their strong sciences) rather than with the more practically oriented academies.

In the years before the First World War, however, agricultural staff began to appeal to the Agriculture Ministry for better provision. Critics had been dismissing the College as an 'asphalt faculty', and staff members complained that the lack of facilities were a source of embarrassment in professional circles.[8] The Ministry seems to have taken this message on board. When Kurt von Ruemker was considering a call to the College in 1912, for example, he made it a condition of his acceptance of the newly established chair of agronomy that the College would establish an experimental field under his direction. The Ministry agreed to his request, probably because about the same time it had anyway decided to move several institutes out to Potsdam where they were to have proper experimental grounds.[9] By the outbreak of war, however, nothing had been accomplished, and after the war plans got underway slowly, angering von Ruemker who resigned in 1919 when the Ministry was unable to reassure him. Replacing him proved difficult because of the continuing lack of experimental fields. Eventually in 1926 a new institute with experimental grounds was completed at Dahlem, though von Ruemker's successor felt that it was still far too small.[10]

With such poor facilities for its agricultural disciplines, the College could hardly claim to have enjoyed the respect of the farming community. To be fair, it had offered short courses for working

8 Skibbe, 'Zur Vorgeschichte'; Delbrueck, *Die Koenigliche Landwirtschaftliche Hochschule*. For complaints in 1914 and 1918 see GStaA, Rep 87B, Nr. 20058.

9 There was a precedent for this decision; about a decade earlier the Education Ministry had agreed to move the Botanical Garden and various University institutes into the suburb of Dahlem where they, too, would have more space (Schiera, *Laboratorium*, 280–282).

10 On Ruemker's anger see GStaA, Rep 87B, Nr. 20088, 20090, 20058. Others were equally disgruntled, among them the geneticist and plant-breeder, Erwin Baur, who repeatedly threatened to leave and eventually departed in 1928 for the Kaiser-Wilhelm-Institute for Plant-Breeding. (GStaA, Rep 87B, Nr 20058; Schiemann, 'Erwin Baur', 78–81). On the continuing lack of fields, see GStaA, Rep 87B, Nr. 20090. On the new institute in 1926 see Opitz, 'Das Institut fuer Acker- u. Pflanzenbau'.

farmers every winter since 1885 which quite often attracted over a hundred students. But few of these courses were actually on mainstream agricultural subjects (eg, meteorology, fisheries technology), and to judge from the attendance figures, those taking part appear to have been more interested in agricultural history or politics than in practical advice. Moreover the College's full-time agriculture students were less likely to come from farming backgrounds than almost any other agricultural institution, and relatively few intended to go into farming.[11]

As far as degree programmes were concerned, Berlin offered the two- and three-year programmes like every university. But it was keen to expand its range with a series of more advanced degree courses in plant-breeding, animal-breeding, agricultural engineering, or food engineering, on the basis of which students could either enter specialised employment or go on for a doctorate. In 1912, similarly, teaching in law, administration, insurance and cooperatives was expanded with an eye to developing an eventual degree in agricultural administration.[12] Thus the College evidently hoped to recruit those students interested in agricultural industry and administration rather than in farming.

Right from the start the College's ties to one branch of industry – food and drink – were very strong. Before the College was founded, the city had possessed a number of research institutes set up by this sector, among them a laboratory for the sugar industry (1867) and one for the distilling industry (1874). The industrial associations behind both of these institutes lobbied energetically for state assistance in various forms (while retaining overall control of the institutes' activities), including an institutional link with higher education. As a result

11 On short-courses for farmers *see Landwirtschaftliche Hochschule Berlin: Jahresberichte*; for others see Oldenburg, *Das landwirt. Unterrichtswesen.* Among Berlin's students only 24% came from farm families (vs. 53% at Goettingen); they were instead more likely to be the sons of civil servants than at almost any other institution. Only 22% of Berlin's students intended to go into farming vs. 60% at several other places (*Deutsche Hochschulstatistik*, 1928).

12 20 July 1912, Rektor to Min. Landwirtschaft, GStaA, Rep 87B, Nr. 200153. *Landwirtschaftliche Hochschule Berlin: Jahresberichte*, 1911–12 and 1928–29.

the sugar institute was incorporated into the College at its founding in 1881, and the distillers' institute was taken on board a year later. In 1889 a brewers' association (*Verein Versuchs- und Lehranstalt fuer Brauerei*) contributed to the funding of an experimental brewery at the College. By the 1890s with space at the College becoming tight, the brewers association joined with the Association of German Distillers (*Verein der Spiritusfabrikanten*), the Association of the German Starch Industry (*Verein der Staerke-Interessenten*), and the Ministry of Agriculture to jointly establish an Institute for Fermentation Research and Starch Manufacture (*Lehr- und Versuchsanstalt fuer die Gaerungsgewerbe und Staerkefabrikation*) at the College. The Institute was enormous. Its budget in 1905 was three times the size of the rest of the College's, and it alone employed as many scientists with doctorates as the College possessed academic staff. Soon after it was established, yet another industrially sponsored institute was acquired: in 1899 the Association of German Millers (*Verband Deutscher Mueller*) agreed to partly fund an experiment station at the College.[13]

By 1906 the directors of both the sugar and the fermentation institutes had been given chairs at the College, and by 1913 these two institutes, along with that for cereals-processing, constituted a separate Faculty within the College. A third Faculty was dedicated to surveying and cultivation technology (*Kulturtechnik*).[14] Although nominally a college of agriculture, therefore, Berlin was a much more heterogeneous institution, as was also evident from the composition of its student body. Since Berlin was the only institution in Prussia where students could qualify in distilleries engineering, brewing or sugar manufacture, a substantial proportion of them were registered for these degrees. Over the period from its founding to 1918 one historian estimates that students were about equally split between agriculture

13 In 1907 the experiment station was recast as an independent Institute for Cereals-Processing (*Versuchsanstalt fuer Getreideverarbeitung GmbH*) which was loosely affiliated to the College. On all of these institutes see Wittmack, *Die Koenigliche Landwirt. Hochschule*; Harms, 'Wittmack'; Ruske, *Reichs- und Preuss. Landesanstalten*, 25–27.

14 Wittmack, *Die Koenigliche Landwirt. Hochschule*.

and surveying/food engineering. Toward the end of this period a clear majority were studying agriculture, but as late as 1900–01 less than half of the College's students did so; nearly as many were enrolled in the Faculty of Surveying as in Agriculture.[15]

In view of the generous provision for the natural sciences and the strong presence of other non-agricultural subjects, it comes as no surprise that the agricultural curriculum placed a heavy emphasis upon the basic sciences. To be sure, the first year of Berlin's two-year degree did include agronomy and animal husbandry, but they were taught by lecture only, and the number of weekly hours devoted to them was dwarfed by those in the sciences (8 vs. 41). And although it was a fundamental agricultural subject, farm management was not taught until the second year.[16] By the eve of the First World War, again, there were signs of change; staff agreed that the curriculum was too heavy on the science side and that more hours should be devoted to agriculture. But they clearly had no intention of fundamentally recasting the curriculum in a 'practical' direction; the reduction in science hours, they emphasised, 'should not be too radical nor should utility be the sole criterion' since students at Berlin should get a

> well rounded and science-based education [*naturwissenschaftliche Allgemeinbildung*...] which would enable them to recognise the general laws underlying the facts and processes encountered in practical agriculture.[17]

15 On enrollment by faculty see Oldenburg, *Das landwirt. Unterrichtswesen* and *Landwirtschaftliche Hochschule Berlin: Jahresberichte*, 1900–01 and 1913–14 (figures for the winter semester). The estimate is from Klemm et al., *Von der Koeniglichen Akademie*, 35.

16 *Landwirtschaftliche Hochschule Berlin: Jahresberichte*, 1892–93 through 1913–14 (weekly hours calculated for 1909–10); cf. Wittmack, *Die Koenigliche Landwirt. Hochschule.*

17 The proposed new curriculum allocated roughly equal hours to courses in science and in agriculture (24 June 1912, Rektor to Min. Landwirtschaft, GStaA, Rep 87B, Nr. 20153). During the 1920s, nonetheless, Berlin's curriculum was perceived in agricultural circles as being excessively dominated by the sciences (Theodor Roemer to Min-Direktor Arnoldi, 26 Aug 1930, GStaA, Rep 87B, Nr. 20095).

Apart from those courses offered by the College itself, students were encouraged to attend humanities courses at the University in order to acquire a 'well-rounded education' which would 'protect the student from superficiality and a one-sided attitude to life'.[18] Through their choice of language, it is clear that staff were declaring their allegiance to the ideology of cultivation then dominant at the universities.

But the 'scientific' orientation of the College's teaching is also visible in other ways. Since some of the science courses taken by agricultural students (eg, in physics) were also taken by students of surveying, they were probably not tailored to agricultural problems. The same may have applied to other agricultural courses since some of them were taken by large numbers of students from both the University of Berlin and the College of Veterinary Medicine.[19]

Given this emphasis upon basic sciences and the limited facilities for practical demonstrations before the First World War, one might have expected the College to appoint staff with rather 'academic' backgrounds and little practical experience. This does not, however, seem to have been the case. Biographical information on professors of agriculture appointed prior to 1906 suggests that many of them had practical experience at the time of their appointments, and even some of the science professors chose to work on agricultural topics. From about 1912 through the 1920s, too, the minutes of appointments proceedings indicate quite clearly that the College was seeking to appoint individuals who possessed substantial practical credentials or interests. And in appointing such figures as von Ruemker, Aereboe,

18 From a lecture-booklet of 1922–23 cited by Papendieck, 'Die Entwicklung', 43.

19 In 1909–10, for example, 316 students from the College were enrolled in its agricultural courses alongside 332 from the University and the Veterinary College. Veterinary students took introductory zoology at the College from 1896 until the 1960s (Herbst, 'Zur Geschichte der Zoologie'). The College's Faculty of Food Engineering, on the other hand, was more concerned about who taught its students the sciences and how. Since all of these courses were taught by staff from that Faculty, they were probably also tailored to the problems of food engineering, possibly because of the close involvement of various food industry associations in the Faculty *(Landwirtschaftliche Hochschule Berlin: Jahresberichte*, 1909–10, 1913–14 and 1920–21).

Johannes Hansen and Kurt Opitz they clearly succeeded.[20] This might seem to indicate that despite weak links to the farming community, a science-oriented curriculum, poor demonstration facilities, and Ministerial indifference, Berlin's ethos had begun by the eve of the First World War to shift in a more practical direction.

The evidence, however, suggests otherwise. There are numerous indications that staff were more concerned with the College's academic status than with its standing in agricultural circles, and this was to have important consequences for teaching and research. For one thing, staff were remarkably preoccupied with the external trappings of prestige. In 1912, for example, they sought the right to wear gowns (*Amtstracht*) at all academic ceremonies because, as they observed, this was the custom at 'the older universities', and its absence would signal fewer rights.[21] Two years later they requested that the designation of their decision-making committee should be changed from the 'body of teachers' (*Lehrerkollegium*) to the 'body of professors' (*Professorenkollegium*) since the former term was generally used at lower level schools rather than at universities. And in 1917 they requested that appointments documents should use the term *ordentlicher Professor* since it was the term used at all German universities and had recently been adopted at Prussian engineering colleges.[22]

Various attempts by the Ministry to prod the College in a more practical direction met with resistance. In 1925, for example, the Ministry proposed a new draft of the College's statutes which read:

> The Agricultural College should provide for academic instruction and research [...] with particular reference to the practical needs [of agriculture]. In addition

20 On earlier appointments see Wittmack, *Die Koenigliche Landwirt. Hochschule*. On filling various chairs 1912–1914, see GStaA, Rep 89H, Nr. 31930 and Rep 87B, Nr. 20087. On the chair of agronomy in 1920 see GStaA, Rep 87B, Nr. 20091. On the chair of botany in 1932 see GStaA, Rep 87B, Nr. 20096. On the chair of animal breeding in 1928 see GStaA, Rep 87B, Nr. 20094. Mommsen, *Geheimrat Hansen*; Tamm, 'Kurt Opitz'; Opitz, 'Acker- und Pflanzenbau'.

21 21 May 1912, Rektor to Min. Landwirtschaft, GStaA, Rep 87B, Nr. 20088.

22 Letters of 10 July 1914 and 27 June 1917, Rektor und Professorenrat to Min. Landwirtschaft, GStaA, Rep 87B, Nr. 20058.

to students' technical training, it is also responsible for their general [...] character-formation [...].

Staff objected, however, to the phrase 'with particular reference to the practical needs [of agriculture]' and requested that in the following sentence 'scientific training' (*wissenschaftliche Ausbildung*) be used instead of 'technical training' because 'this better expresses the academic character of the College and above all it avoids giving the impression that we are merely a vocational college.'[23]

In a similar vein, staff were keen to see the College expand but in ways which enhanced its academic status. In 1914, for example, they asked the Minister for permission to establish new Faculties of Science and of Social Sciences. The former was to be authorised to award a doctorate in applied sciences, and this was thought to be particularly important because it would 'reflect our tasks of promoting pure and applied sciences.'[24] Another route for expansion which had the staff's support was to have the Forestry Academy at Hann-Muenden incorporated into the College as a separate faculty. There was a move underway at the moment, they argued, to amalgamate small vocational colleges with larger units: the veterinary colleges in Dresden and Munich were to be merged with the universities in Leipzig and Munich, and Berlin's mining academy was to be become

23 First quote: 'Die Landwirt. Hochschule soll dem wissenschaftlichen Unterricht und der Forschung [...] unter besonderer Beruecksichtigung der praktischen Beduerfnisse dienen. Sie hat hierbei neben der fachwissenschaftlichen Ausbildung auch der allgemeinen [...] Charakterbildung der Studierenden Rechnung zu tragen [...].' The second: '[...] hierdurch der wissenschaftliche Charakter der Hochschule staerker zum Ausdruck kommt und dass vor allen Dingen der Eindruck vermieden wird, dass es sich um eine reine Fachschule handle.' Both quotes are from 5 Mar. 1925, Hansen to Min. Landwirtschaft, GStaA, Rep 87B, Nr. 20058.

24 10 July 1914, Rektor und Kollegium to Min. Landwirtschaft, GStaA, Rep 87B, Nr. 20058. The timing of this statement may have been significant. For a month earlier at celebrations of the 50th anniversary of Halle's Agricultural Institute its director had declared that the academies and colleges were just 'higher vocational institutions' (*hoehere Fachschulen*) which could not match the universities in their provision for the sciences (Wohltmann, 'Die Festfeier des 50-jaehrigen Bestehens').

part of the city's engineering college. It was thus essential, they stressed, that the Agricultural College should be free to grow in order that it did not 'sink back to the level of an agricultural academy or vocational college'.[25]

Despite periodic claims that they sought to maintain close ties to practical agriculture,[26] Berlin's staff were in fact quite concerned to secure greater institutional autonomy. In 1916 and again in 1921–22, for example, they requested that the Minister dissolve the College's board of overseers (*Kuratorium*) on the grounds that no other college or university in Berlin had one. That the College alone possessed such a board was 'likely to harm its reputation'. Their request was granted.[27] After the war, when staff got wind of the Ministry's plan to introduce advisory committees (*Beiraete*) at both Prussian agricultural colleges, on which agricultural interests would be represented, they reacted sharply. The aim of such committees, they argued, was to link science and practice, but College staff already had personal contacts with agricultural organisations and representatives of the farm community and always kept an eye on practical issues in their teaching. The 'danger' of such a committee was that its agriculturalist members might fail to appreciate science's need for freedom and would insist upon too restricted a conception of science's obligations to practice. Such a committee, they concluded, might be all right for a vocational school but not for a scholarly institution (*wissenschaftliche Hochschule*).[28]

That Berlin's staff were so sensitive about matters of status reflects the fact that the College then occupied a lower rung on the academic ladder than both the universities and the engineering colleges and was obsessed with achieving parity. One obstacle in their path was that the College's entry requirement fell far short of that

25 17 Mar 1915, Rektor to Min. Landwirtschaft, GStaA, Rep 87B, Nr. 20058.

26 eg, Wittmack, Die Koenigliche Landwirt. Hochschule, 53.

27 13 July 1921, Rektor and Professorenrat to Min. Landwirtschaft; see also 12 May 1916, Rektor and Kollegium to Min. Landwirtschaft, both in GStaA, Rep 87B, Nr. 20071.

28 23 May 1920, Rektor und Professorenrat to Min. Landwirtschaft, GStaA, Rep 87B, Nr. 20058. I have been unable to clarify whether such an advisory body was nonetheless established.

required for University entrance.[29] Moreover the staff's self-esteem was probably dented by the fact that in 1900, when the engineering colleges acquired the right to award their own doctorates (thus gaining a degree of parity with the universities), the College did not. What it did acquire was the right of its students to submit a doctoral dissertation to the *University*. This meant, however, that the College's students had to take doctoral examinations, not just in agriculture, but in three subsidiary subjects: philosophy, a natural science and a social science. Some staff found this dependency demeaning: Kurt von Ruemker deemed it an 'unworthy farce' that his students' dissertations should have to be approved by a professor of philosophy, and he refused on principle to take on any more doctoral candidates. When in 1910 the Veterinary College acquired the right to award its own doctorates, College staff began to press for the same privilege, finally succeeding in 1918.[30]

The pattern of staff career-trajectories points in the same direction. Of those called to chairs at the College, the great majority had previously been employed at veterinary colleges, engineering colleges, or agricultural experiment stations. The only ones who had come from a university were scientists who had previously held only associate professorships (*Extraordinariate*). After arriving at the College, many (especially the science staff) moved on after a few years to a university, engineering college or high level civil service post and only rarely to an agricultural college.[31]

For basic scientists in particular the College functioned as a 'waiting room' until something more attractive came along. This pattern had important consequences since transients were unlikely to make much effort to address agricultural problems in their research or

29 Over the period 1890–1906, for example, only 10 to 20% of the College's students had passed the *Abitur* examinations generally required for University entrance (Wittmack, *Die Koenigliche Landwirt. Hochschule*, 271).

30 On doctoral examinations at the University see *Landwirtschaftliche Hochschule Berlin: Jahresberichte*, 1900–01. On awarding its own doctorate see 5 Nov 1914, von Ruemker to Rektor, GStaA, Rep 87B, Nr. 20058; and 23 July 1918, von Eisenhart-Rothe to Kaiser, GStaA, Rep 89H, Nr. 31930.

31 *Landwirtschaftliche Hochschule Berlin: Jahresberichte*, 1897 to 1921, plus additional instances I have run across.

even in their teaching. Consider the professors of zoology. Arriving in 1905, Ludwig Plate worked on evolution and heredity until leaving in 1909 for a chair at the University of Jena. His successor, Richard Hesse, was primarily interested in functional morphology, ecology and biogeography, leaving in 1914 for a chair at the University of Bonn. The College then offered the chair to Max Hartmann, who was working on sex determination in protozoa at the Kaiser-Wilhelm Institute for Biology, but Hartmann declined. Only in 1915 did they manage to recruit someone who was interested in agriculture: Richard Heymons worked on applied entomology and stayed for twenty years.[32] The botany professors were more inclined to stick around, and at least some of their research was agriculturally relevant, though in his course on plant systematics even one of the more agriculturally oriented professors (Ludwig Wittmack) insisted upon treating the entire plant kingdom rather than focussing upon cultivated plants.[33]

For young academics bent on a university career, furthermore, the College's location made it a relatively desireable waiting room. Quite a few College staff, for example, held adjunct appointments at the University,[34] and the city's vibrant scientific life also extended to the engineering college, Kaiser-Wilhelm Institutes (from 1911) and a variety of other specialised research institutes at whose colloquia one could hope to network and make an impression. And since Berlin was the seat of both Prussian and Reich governments, lobbying civil servants in the relevant ministries was relatively easy.

Indeed, its geographic location undoubtedly had a substantial impact upon the general character of the College. Maintaining ties with the farming community, as various staff pointed out, was a good deal more difficult in Berlin than in a small college town situated in an

32 Herbst, 'Zur Geschichte der Zoologie'; 28 Mar 1914, Max Hartmann to A. Harnack, Emil Fischer papers.

33 Natho and Wiedenroth, 'Zur Geschichte der Botanik'.

34 Leopold Kny, professor of botany, was also associate professor at the University. Ludwig Wittmack, professor of agricultural botany, was also associate professor for systematic botany at the University. Ludwig Plate, professor of zoology, was also in 1905 a Privatdozent at the University, as was the professor of physics, R. Boernstein (Wittmack, *Die Koenigliche Landwirt. Hochschule*).

agricultural region.[35] Far more strongly represented in the city was the food and drinks industry which, as we saw, was a major player at the College. Just as important as the College's location in relation to the local agricultural economy, however, was its proximity to other higher education institutions. Situated in the Invalidenstrasse in central Berlin, it was only a few blocks away from the University on Unter den Linden. That offered the advantage that agriculture students could easily attend lectures at the University (a fact which was often played up in distinguishing the College from the 'isolated' academies at Hohenheim and Weihenstephan). And it also allowed the College to boost its numbers by attracting students from the University and other institutions, and high student numbers guaranteed resources as well as the distinction of being the largest of the higher agricultural institutions in the decade before the First World War. On the other hand proximity also brought disadvantages, for it meant that staff could not avoid being reminded that in terms of the academic status hierarchy they were well down the ladder.

Thus Berlin's pronounced 'scientific' orientation was the product of its staff's unquenchable thirst for academic status, the Ministry's disinclination to push for reform, and the local economy's acquiescence. Endowed with exceptionally poor demonstration facilities – aggravated by the Prussian Agriculture Ministry's failure to remedy this before the First World War – the College was in no position to compete with the academies as a 'practical' institution. On the other hand, as a vocational institution it could hardly compete with the universities in terms of status; its best hope was to achieve parity of esteem with the engineering colleges. Caught in the middle of the academic hierarchy, therefore, the College played to its strengths and sought to maintain a middle course. Exploiting its advantages over the academies – a wider range of subjects organised in several faculties, strong sciences, its cosmopolitan location and its college-status – it cast aspersions upon 'mere vocational colleges' while emulating the

35 Settegast, *Erlebtes*, 287; Opitz, 'Das Institut fuer Acker- und Pflanzenbau'. Even those who admired the College's accomplishments had to concede that Berlin was the 'worst imaginable location' for an agricultural college (Remy, 'Die landwirt. Akademie in Bonn').

universities wherever possible. In short, Berlin was the inveterate social climber.

The Agricultural Institute at the University of Halle

Halle was the oldest of the university institutes and had the largest number of students until just after the turn of the century (when it was overtaken by Berlin). It also had the highest international profile among German agricultural institutions, attracting large numbers of students from abroad (mainly from Eastern Europe). In the decade before the First World War, for example, between 30 and 40% of Halle's students were from outside Germany, compared to 10 to 20% at other Prussian institutions.[36]

Just why Halle possessed such international appeal is not clear, though it probably derived in part from Kuehn's reputation as a pioneer of university education in agriculture. It is unlikely to have been due to the Institute's demonstration facilities. To be sure, the 115 hectar experimental field, just a 10 minute walk from the university, made Halle one of the better equipped of the university institutes, but in other respects it did not stand out. The University did not have a model farm (unlike Goettingen or the academies), partly because Kuehn felt it would take too much time and energy away from his scholarly work. Nor did it possess as many agricultural institutes and professors as Breslau or Bonn. And its state funding in 1910 was less than half of Berlin's. As late as 1912 Halle – in contrast with most other universities and academies – still had no laboratory facilities of its own for agricultural chemistry, having relied since the 1860s on the nearby facilities of the Chamber of Agriculture's experiment station. Moreover its lack of a chair dedicated to farm management (until

36 On student numbers see Hansen, *Das landwirt. Unterrichtswesen* (1920), 19. On foreign students see Oldenburg, *Das landwirt. Unterrichtswesen; Landwirt. Jahrbuecher*, 1910; *Kuehn-Archiv*, Vol. 5 (1914), xvi and appendix.

1914) placed it behind, not only all of the colleges, but also several universities.[37]

Nevertheless, Halle seems to have enjoyed good ties to the region's large landowners and agricultural industry. The Prussian province of Saxony enjoyed exceptionally fertile soil, and its large farmers were relatively prosperous as well as keen on new technology. As early as 1837, for example, a group of regional estate-owners interested in the advancement of agricultural knowledge had established a national agricultural society (*Versammlung deutscher Land- und Forstwirte*), and in 1851 the first German agricultural experiment station was established at Moeckern.[38] And it was partly this regional culture of innovation which had attracted Julius Kuehn to Halle. The province was also a centre for various branches of industry (fertiliser, sugar and plant-breeding) which seem to have taken an interest in the Institute, as did estate-owners. When regional sugar-beet growers, for example, approached Kuehn for help in dealing with a pest outbreak, he was able to demonstrate that the soil was infested with nematodes.[39] While Erich von Tschermak was working in the province as a student-apprentice in the early 1890s numerous landowners recommended that he enroll at Halle, and at its 50th anniversary in 1914 donations to the Institute's jubilee fund were made by several fertiliser companies and four large landowners (who were duly rewarded with honorary doctorates). Given Halle's location, therefore, it is perhaps not surprising that a much higher proportion of its students than Berlin's intended to go into farming (63% vs. 22%).[40]

37 On Kuehn's view of model farms see Wohltmann, 'Das landwirt. Institut', 5. On facilities see *Landwirtschaftliche Jahrbuecher*, 1910 and *Mentzel & Lengerkes Schreibkalender*, 1913. On budgets for 1910 see Oldenburg, *Das landwirt. Unterrichtswesen*. On agricultural chemistry see 21 Mar 1912, Wilhelm Schneidewind to Kultusminister, Personalakte-Schneidewind, UA-Halle. On farm management see *Mentzel & Lengerkes Schreibkalender*, 1913; Titze, *Datenhandbuch*; and *Chronik der Universitaet Halle* (1916–26), 207–208.

38 Finlay, 'Science, Practice and Politics', 46ff.

39 Mueller and Klemm, *Im Dienste der Ceres*, 211–212.

40 On Kuehn's attitude to Halle see Hansen, *Das landwirt. Unterrichtswesen* (1919), 12; Schwabe, 'Zur Idee', 43; Dittrich, 'Julius Kuehn', 63. Tschermak, *Leben und Wirken*, 28. In the 1890s the landowner and eminent plant-breeder,

That does not mean, of course, that Halle's curriculum was 'practically' oriented. On the contrary; as we saw in Chapter Three, Julius Kuehn promoted a conception of agriculture as a form of applied biology and required doctoral students to select problems which could be solved through laboratory work. (Significantly, his original wish had been that his new institute would be named an 'agricultural-physiological' institute.[41]) In keeping with the ritual invocations of pure science so common in late 19th century German universities, he decried the 'selfish' and profit-seeking attitudes fostered by 'one-sided technical education' (*einseitige Fachbildung*), advocating instead the 'broader and higher' perspective conferred by studying at a university. Just as in medical education, Kuehn noted, this broader perspective was conveyed in part by the fact that the student began by studying the natural sciences – which were taught without any reference to agriculture – in order to provide the essential foundation for the subsequent study of agricultural subjects. The farmer was not merely a 'man of commerce' (*Gewerbetreibender*), Kuehn stressed; he was also a representative of his profession and an educated man. Therefore, it was essential that the student also attended courses, not only on economics, law and politics, but also on literature, history and philosophy.[42] (Clearly the 'farmers' Kuehn had in mind were estate-owners rather than peasants.) Before the First World War, therefore, Kuehn took care to steer Halle's curriculum away from anything 'practical'.

But why? Part of the explanation is probably the Institute's location within the Philosophical Faculty. As we saw in Chapter Three, relations with the rest of the Faculty were strained in certain respects, aggravated by the fact that the entrance requirement to study agriculture was lower than for almost any other subject in the Faculty. During the 1890s, accordingly, agriculture students were regarded by the science professors as somewhat inferior, a taint which Kuehn

Wilhelm Rimpau, was also awarded an honorary doctorate by the University (Krzymowski, *Geschichte*, 271). On donations see *Kuehn-Archiv*, Vol. 6 (1915), 14–16. On students' career aims see *Deutsche Hochschulstatistik*, 1928.

41 Dittrich, 'Julius Kuehn', 65.

42 Kuehn, Programm fuer das Studium, 18.

sought to dispel by emphasising his doctoral students' experimental skill. For their part, his philologist colleagues were not persuaded that agriculture belonged in a university at all. Another disadvantage of this location was the fact that the Institute's doctoral candidates were obliged to pass exams in a science and philosophy as subsidiary subjects. And Kuehn's successor reckoned that Faculty resistance was one of the reasons why at the outset Kuehn had requested only an experimental *field* for the Institute rather than a model farm. However, unlike his counterparts elsewhere (see below on Breslau), Kuehn did not try to carve out more elbow-room for the Institute. He not only defended Philosophy as the proper site for his institute but also resisted proposals for a separate 'Agricultural Faculty' for fear that its staff might then be ridiculed in the University as 'manure-men' (*Mistiker*). The study of agriculture at Halle, he emphasised to prospective students, was situated 'in the midst of the University just like the study of any other science', and the Agricultural Institute was an integral part of the University, 'just like the natural science and medical institutes'.[43]

On the eve of the First World War, however, Halle's orientation toward science/practice began to shift. In 1909 Ferdinand Wohltmann (1857–1919), whom Kuehn (then 80 years old) had brought to Halle four years earlier to share his workload, formally became director of the Institute. Never the ivory-tower scholar, Wohltmann had a strongly economic perspective on agriculture which distinguished him fundamentally from Kuehn. The son of an estate administrator, he had worked for six years on a farm before completing his secondary education. Before coming to Halle he had developed especially close ties to farmers in both Breslau (1892–94) and Bonn (1894–05). Armed with this experience, Wohltmann's vision of the agricultural sciences was much broader than Kuehn's. Academics in this field, he wrote,

43 On experimental skill see Tschermak, 'Ein Leben fuer die Zuechtung'. On philologists see Conrad, *Lebenserinnerungen*, 153–156. On doctoral exams see Tschermak, *Leben und Wirken*, 30–32. On field rather than farm see Wohltmann, 'Das landwirt. Institut', 5. On manure-men see Dittrich, 'Julius Kuehn', 65. The quote is from Kuehn, *Programm fuer das Studium*, 3.

required not only knowledge of the natural sciences but also of economics, as well as first-hand experience of farming.[44]

Immediately following Kuehn's death in 1910, therefore, Wohltmann began to implement plans for change which he had been harboring for some time.[45] What he had inherited from the 'Meister' was an Institute poorly equipped in several areas. But between 1910 and Halle's 50th anniversary in 1913, he initiated a wave of expansion which rapidly improved the range of Halle's facilities. Among these were new buildings for agronomy, a plant-breeding station, a separate chair for animal-breeding, and improved facilities, an assistant, and the prospect of a separate chair for farm management. The new developments were very welcome in agricultural circles. Prussian Saxony's Chamber of Agriculture, for example, had been concerned at the poor provision for animal-breeding over the previous decade and was thus pleased with Wohltmann's plans to improve provision for it as well as with the new professor for the subject (Hermann Nathusius). And by the eve of the First World War the Prussian Education Ministry reckoned that Halle had become the top agricultural institution in Germany.[46]

44 Wohltmann, 'Die Festfeier des 50-jaehrigen Bestehens', 11; see also Schwabe, 'Zur Idee', 59–60. On Wohltmann see Disselhorst, 'Gedaechtnisrede'; Kalt, 'Ferdinand Wohltmann'; Kattenbusch, 'Zum Gedaechtnis'; Anon. [obituary of F. Wohltmann]; and Wolff, 'Ferdinand Wohltmann'.

45 Only 2 weeks after Kuehn's death, Wohltmann sent a discussion paper (with 9 appendices) to the Education Minister, outlining essential new facilities (14 Sept 1910, Kultusminister to Min. Finanz, GStaA, Rep 151/IC, Nr. 6829).

46 On farm management see the preface to Wohltmann, *Festschrift*, xiii–xiv; and 26 Aug 1910, Kultusmin. to Min. Finanz, and 20 Juni 1911, copy of Wohltmann to Kurator, both in GStaA, Rep 151/IC, Nr. 6829. On animal-breeding see 31 Oct 1912, copy of Graf von der Schulenberg to Kultusmin., GStaA, Rep 151/IC, Nr. 6829. Nathusius's successor in 1915, Gustav Froelich, also had close ties to local animal-breeders (6 June 1923, copy of Landwirt. Kammer fuer die Provinz Sachsen to Kultusmin., GStaA, Rep 151/IC, Nr. 6829). After the First World War Saxony's cattle-traders took a close interest in developments at Halle and made a substantial investment in improved facilities for animal-breeding (7 May 1919, K. Haenisch to Min. Finanz and 29 Dec 1919, Kultusmin. to Min. Finanz). The Education Ministry's assessment is in 23 Feb 1914, Kultusmin. to Min. Finanz, all three in GStaA, Rep 151/IC, Nr. 6829.

Securely established at the top of the academic status hierarchy, therefore, Halle was well placed to resist any pressure which the region's farmers (or the Ministry) might have wanted to exert. But such pressure is little evident, probably because the Institute's 'scientific' approach to agriculture found support among both the region's technically progressive farmers and its agricultural industries. High academic status was a valuable resource, allowing Halle to recruit academically qualified staff and large numbers of students from abroad, and thus had to be maintained in good condition. This is why Julius Kuehn took care not to make waves within the Philosophical Faculty: agriculture was deemed to be an applied science rather than a commercial subject; doctoral students were to select laboratory-based problems; and farm management was assigned a low priority. Orienting itself toward 'science' was thus Halle's best strategy for securing legitimacy since it consolidated the Institute's position at the top of the status hierarchy while not alienating powerful groups in the regional economy. Kuehn's death, however, allowed his successor, guided by a more practical vision of agricultural education, to mobilise ministerial support and set the Institute on a new course.

The Agricultural College at Bonn

The College[47] at Bonn was a more practically oriented institution than either Berlin or Halle. Like Berlin, it originated in 1819 as an institute within the newly founded University of Bonn (1818), but in 1846, following a longer period of stagnation and mixed fortunes, it was developed – at the request of the Agricultural Association of Rhein-Prussia – into an independent institution which was to be closely associated with the University. (This association made it easier for

47 Like the institutions at Hohenheim and Weihenstephan, Bonn was originally established as an 'academy' and only much later (in 1919) upgraded to a 'college'.

Bonn to withstand Liebig's criticism of the academies in the 1860s.)[48] By 1900 the College was an important institution because of its size. In the years before the First World War it was second only to Berlin and Halle in student numbers, enjoyed a state subsidy about equal to Halle's, and was second only to Berlin in the number of its institutes (ten).[49]

For comparative purposes, Bonn is an interesting institution because although its geographic location was radically different from Berlin's, the two colleges were quite similar in formal structure and status. Like Berlin, Bonn was overseen by the Prussian Ministry of Agriculture, and it possessed separate faculties for Agriculture and for Surveying & Cultivation Technology.[50] Bonn often described itself as 'closely associated' (*eng verbunden*) with the University, no doubt in part to distinguish itself from the 'isolated' academies. But in formal terms this seems to have consisted in nothing more than the fact that its students were enrolled in the University's Philosophy Faculty. To be sure, when Theodor von der Goltz took over as director of the College from 1896, he sought closer ties to the University and himself took up a new chair of agriculture at the latter. But double appointments of this kind did not become the norm. In 1900, for example, apart from von der Goltz, only one other of the 12 staff at the College also held an appointment at the University, and in 1918 only 3 of them did so. Conversely before the First World War only about one-quarter of the adjunct lecturers (*Honorar Dozenten*) at the College were University staff. And while staff at the College may have been favourably disposed towards their brethren at the University, the feeling was evidently not always mutual. From 1892 the College's professors enjoyed formal equality with professors at the University,

48 Kick, 'Kurze Geschichte'.

49 Bonn had about 200–300 students versus 300–400 at Halle and somewhat more at Berlin (*Landwirt. Jahrbuecher*, 1910; Hansen, *Das landwirt. Unterrichtswesen* (1920), 19). On subsidies see Oldenburg, *Der Ausbau*, 110. On institutes see Oldenburg, *Das landwirt. Unterrichtswesen.*

50 As at Berlin, agriculture students were outnumbered by surveyors (by nearly 10 to one) at the end of the 19th century, though by the First World War they had grown to a substantial majority (*Bonn-Jahresberichte*, 1909–1914).

but they were not paid as well, and differences of social status persisted for some time.[51]

Despite these structural similarities with Berlin, however, Bonn had a much stronger practical orientation. For one thing, it was much better equipped with experimental grounds and was one of the few institutions to possess a model farm (125 ha). In terms of curriculum students were confronted with agricultural issues from the start, taking courses in agronomy and animal husbandry in their first semester as well as attending demonstrations at the model farm.[52] As at most agricultural institutions in the late 19th century, farm management at Bonn was initially taught by staff whose principal interests lay elsewhere. But with the arrival of von der Goltz in 1896, Bonn acquired a specialist in this field (nearly 20 years before Berlin). A dedicated chair and institute were created in 1906 (about the same time as in Breslau and second only to Hohenheim and Weihenstephan), making Bonn relatively advanced in its provision for this subject. Through von der Goltz, as well as his successors, Friedrich Aereboe (briefly) and Theodor Brinkmann (1911–1948), Bonn enjoyed a reputation as a centre for innovative work in farm management.[53]

As far as provision for science teaching was concerned, one might have expected the College to require its students to attend science courses at the University since they were anyway enrolled there, and university courses enjoyed prestige. Significantly, however, about 1860 the College decided (and the Ministry agreed) to appoint its own science staff in order that they could focus upon those areas of the

51 On double appointments see *Die Koenigliche Preuss. Landwirt. Akademie Bonn: Ratschlaege*, 16–17; *Chronik der Rheinischen; Bonn – Jahresberichte*, 1910 and 1913. On social inequalities see Kick, 'Kurze Geschichte' and Remy, 'Die landwirt. Akademie in Bonn'.

52 On demonstration facilities see Mentzel und Lengerkes Schreibkalender, 1913; Oldenburg, Das landwirt. Unterrichtswesen. On the curriculum see *Koenigliche Preuss. Landwirt. Akademie Bonn: Ratschlaege*.

53 On the provision for farm management see Brinkmann, 'Die landwirt. Betriebslehre'. Data assembled from *Mentzel und Lengerkes Schreibkalender*, 1902–1930; Titze, *Datenhandbuch*; Oldenburg, *Das landwirt. Unterrichtswesen*; and *Landwirt. Jahrbuecher*, 1903 and 1910. On Bonn's reputation in this field see Frauendorfer, *Ideengeschichte*, 43ff.

sciences which were especially important for farming. Thus by the 1870s there were posts for chemistry, physics, botany, technology, and zoology/mineralogy, as well as institutes for the first three. The College's director around the turn of the century felt it was right that his science staff should tailor their teaching to the needs of agriculture. In no way, he emphasised, did this represent a retreat from the high ideals which each scholar should adopt; it was merely a form of specialisation, comparable to that occurring in all areas of scholarship in the latter 19th century.[54] Unlike at Berlin or Halle, therefore, at Bonn 'tailoring' was regarded as desireable.

In terms of the agricultural economy, the College's location in the Rheinland meant that, in contrast with Berlin (in Brandenburg) or Halle (in Prussian Saxony), nearly all farms in the region were very small. (Table 2.1 shows that Brandenburg, for example, had over ten times the proportion of farms larger than 100 ha and over 400 times as many farms over 500 ha.) Estate-owners were thus a relatively unimportant clientele for the College. Since the city of Bonn was a regional centre, various Rhenish agricultural organisations had established research facilities there with which the College had connections, among them the Agricultural Association for the Rhein Province (the director of whose experiment station was an adjunct lecturer at the College) and the Bee Breeders' Association (whose facilities were made available for teaching). In addition, various research projects at the model farm were undertaken in response to requests from the Ministry of Agriculture and the Chamber of Agriculture or in collaboration with the German Agricultural Society. In view of these facilities and its rural location, it is not surprising that compared to Berlin, more of Bonn's students came from farming backgrounds, and twice as many wanted to go into farming following graduation.[55]

54 On the science chairs see Koernicke, 'Die Landwirtschaftliche Hochschule' and von der Goltz et al., *Festschrift*, 95. On von der Goltz's views see ibid., 98.

55 On connections with associations see *Bonn-Jahresbericht* , 1913. On projects at the model farm see *Bonn-Jahresberichte*, 1911–1914. On student backgrounds and careers see *Die Koenigliche Preuss. Landwirt. Akademie Bonn: Ratschlaege*, 21, and *Deutsche Hochschulstatistik*, 1928.

In common with the other colleges, Bonn underwent various forms of upgrading after the First World War. Like Berlin and the others, Bonn acquired the right to award doctorates in 1919 (though only to candidates who had the *Abitur*). And the following year a small but significant change was made to the College's statutes. Bonn was not only to further teaching and research in agriculture, surveying and the associated sciences, but also to deepen its students' 'general cultivation' and prepare them for citizenship, therewith taking on the high status educational tasks of the universities. This upgrade seems to have been important to staff since it was claimed that their new rights would demonstrate to the outside world that Bonn 'had been added to the ranks of the first-class independent institutions'.[56]

But while they welcomed the hallmarks of formal parity with the universities, there is no evidence that most Bonn staff actually endorsed the anti-utilitarian ideology so common in the universities at that time. As a specialist in farm management, for example, Goltz's view of the nature of the agricultural sciences was much closer to Albrecht Thaer's than to Julius Kuehn's. To regard agriculture as 'applied biology', he argued, was fundamentally mistaken. While the natural sciences provided an essential foundation, the study of agriculture was in its essence a branch of economics, and the task of agricultural scientists was to establish the most economically effective ways in which the natural sciences could be implemented under particular farming conditions.[57]

Furthermore Goltz boasted that his staff at Bonn were strong not just on the technical side but also on the economic one, and the evidence seems to bear him out. That was certainly true for Friedrich Aereboe who was appointed professor of farm management in 1906 (see Chapter Two). It also applied to Johannes Hansen, professor of animal husbandry at the College from 1901 until 1910. Before entering academe, Hansen had been director of an agricultural school for eight years and an estate-administrator for four. Very much a practic-

56 *Bonn-Jahresbericht*, 1919, 4. On the doctorate see Kick, 'Kurze Geschichte'. For the change in statutes compare *Satzungen der Koeniglichen Landwirt. Akademie [Hochschule] Bonn* for 1900 and 1920.

57 von der Goltz et al., *Festschrift*, 95–99.

ally oriented academic, he cultivated close ties to the farming community in the Rheinland and actively promoted cattle breeding and milk-testing associations (*Milchkontrollvereine*) through advisory work and publications.[58] Another instance was Friedrich Wohltmann, professor of agronomy from 1894–1905 and, as we saw above in relation to Halle, anything but an ivory-tower scholar. During his years at Bonn he was instrumental in getting the Rhenish Seed-Multipliers Association (Saatbauverein) founded, was elected chair of the seed-certifying body of a landowners' association (*Original-Saatgut-Kommission* of the *Bund der Landwirte*), and his institute held short courses for Rhenish plant-breeders.[59]

Wohltmann's successor in 1905 was Theodor Remy (1868–1946) who grew up on a Rhenish farm and worked there for about six years after leaving school. Later as an academic Remy worked closely with farmers' organisations in the Rheinland and was instrumental in getting several growers associations established and providing advice to numerous others. When Bonn was upgraded after the First World War to a college with Remy as its first rector, he recommended to the Ministry – as Berlin's staff did not – that it establish an agricultural advisory body (*Beirat*) to facilitate closer ties with the farming community. The advisory body was in fact established in 1920, consisting of three staff from the College as well as eleven farmers (large and small) and representatives of agricultural organisations. Its task was to express a view on educational matters and to suggest new activities which might be developed at the College.[60]

Theodor Brinkmann (1877–1951) was professor of farm management from 1910. The son of peasants, after leaving school he worked on the family farm for six years before opting for higher education. Through his farming experience, he had become aware of the challenge facing small farmers who increasingly needed to be able to compete on world markets. Accordingly, both his doctoral dissertation

58 For Goltz's view of staff see Brinkmann, 'Die landwirt. Betriebslehre'. On Hansen see Mommsen, *Geheimrat Hansen*.

59 See the obituaries of Wohltmann cited in the section on Halle.

60 On Remy see Klapp, 'Theodor Remy' and Anon., *Geheimrat Th. Remy*. On the advisory body see *Bonn-Jahresbericht*, 1920–21.

and his *Habilitation* thesis studied the reasons why Danish agriculture had managed to do this. The key to Danish success, he concluded, was that compared to German cooperatives, Danish ones had taken on a wider range of activities. In much of his subsequent work he sought to identify the obstacles hindering a similar development in Germany. With Aereboe he was generally recognised internationally as a major innovator in the field because he moved beyond the largely empirical and statistical studies of the 19th century to conceptualise the farm as a dynamic system and to formulate the relations among its variables in a causal theory.[61]

In conclusion, though Bonn was formally similar to Berlin in a number of respects, its practical orientation was far stronger. Despite its formal connections to the local university, Bonn was further down the status hierarchy than Berlin (not acquiring college-status until after the First World War) and thus had less to fight for on that terrain. Its advantages lay elsewhere. Bonn's geographic location made it much easier to develop strong ties to farming, and as a small city in the midst of an agricultural region, it provided a natural base for various agricultural organisations with whom the College could collaborate. Furthermore, endowed by the Ministry with excellent demonstration facilities as well as with a staff keen to undertake extension work, Bonn was able to win the confidence of the peasant farmers who dominated the Rheinland's agriculture and who were more likely to favour 'practical' education.

The Agricultural Institute at the University of Breslau

Breslau was neither the largest of the agricultural institutions nor did it enjoy Halle's international reputation. It was, however, a major site of institutional innovation from the 1890s, largely through the efforts of Kurt von Ruemker. It is particularly illuminating to compare its development with Halle's since although von Ruemker, too, faced op-

61 Rintelen, 'Th. Brinkmann'; Andreae, 'Theodor Brinkmann'.

position from the Philosophical Faculty in which his institute was located, he chose to fight back (unlike Julius Kuehn), aided by support from the Education Ministry, and succeeded in turning Breslau into a strong and practically-oriented institution.

In the late 1860s, responding to the debate triggered by Liebig, discussions were underway to close the Agricultural Academy in Proskau (in the Prussian province of Silesia). This closure finally took place in 1881, and in its place the Prussian Ministry of Education established a chair of agriculture in the University's Faculty of Philosophy, along with four associate professorships for various agricultural specialties (all but one of them filled initially by personnel from the former Academy or other local institutions). Within a decade, however, student numbers dropped to less than a third of those at the Academy, in part because Silesian farmers recognised that the University's facilities were markedly inferior to those of the former Academy. Attempting to compensate for this, the Institute sought closer ties to the local Chamber of Agriculture, but this was opposed by the Faculty of Philosophy, prompting several of the agricultural professors in 1890 to tender their resignations.[62]

At that point an *ad hoc* committee, consisting of the remaining academic staff along with representatives from relevant ministries and the local Agricultural Association (*Landwirt. Zentralverein fuer Schlesien*), set out to reorganise the provision for agricultural education, though without consulting the Philosophical Faculty. Angered, especially at the inclusion of the Agricultural Association in the planning process, the Faculty then requested that the new Institute for Agriculture be removed from the University altogether. The Education Ministry chose to ignore this request, and the reorganisation went ahead as planned, though relations between the agricultural staff and the rest of the Faculty remained fraught. When the professor of agriculture later asked for the Faculty's support in expanding the

62 On the new developments of 1881 see von Ruemker, 'Programm fuer das Studium'. The four specialties were agricultural chemistry (*Tierchemie*), animal nutrition, agricultural technology, and veterinary medicine. On the subsequent decline see Zorn, *Die Geschichte*.

provision for Agriculture, for example, it was declined on grounds of the Institute's continuing ties to the Agricultural Association.[63]

As a temporary measure in order to help out, the Chamber of Agriculture and the local Agricultural Association leased a piece of land which was put at the disposal of the Institute as an experimental grounds. By 1895, however, the lease had run out, and the two organisations were exerting pressure upon the Education Ministry, not only to purchase such grounds but to improve provision for teaching and research in a number of disciplines. Arriving that year as the newly appointed professor of agriculture, Kurt von Ruemker found the Institute in a sorry state. Unless he could invigorate it, the Minister told him, the Institute was in danger of being closed.[64]

Ruemker moved quickly. A firm believer in the necessity of specialisation (cf. Chapter Three), within three years he had replaced the original Institute for Agriculture with four separate ones – for plant production, animal husbandry, cultivation technology, and agricultural chemistry-with-bacteriology, each with its own professor and budget. (This was a highly radical innovation at that time, at least for a university; specialisation on this scale in the agricultural sciences would not be introduced at most other universities for another twenty years.) Moreover, von Ruemker chose not to retain overall direction of these institutes himself but instead to introduce a system whereby the chairmanship rotated among the professors. A collegial arrangement of this kind was also unique in the agricultural sciences at that time, both in universities where one professor ruled the roost and in the academies where the professors were subordinate to the director. By 1899 Breslau had also acquired good-sized experimental grounds (33 ha) situated a 15 minute ride from the University. Despite these improvements, Halle still had larger grounds, many more students and twice the level of state funding. But the reforms which von Ruemker

63 Holdefleiss, 'Das landwirt. Studium'.

64 23 Feb 1895, reprint of Stenographische Abhandlungen des Hauses der Abgeordneten, GStaA, Rep 76, Va, Sekt 4, tit X, Nr. 62, Bd 4; von Ruemker, preface to his *Methoden*.

introduced meant that by the turn of the century Breslau had more institutes and professors than any other university including Halle.[65]

To some extent the fact that institutional innovation in agriculture occurred so early at Breslau may have been part of a more general pattern there. For it was also at Breslau that the first physiological institute (1838) in Germany was established, and in 1853 Breslau was the first Prussian university to get a chemical institute which measured up to Liebig's standards. Moreover the institute of plant physiology created for Ferdinand Cohn in 1866 was the first such institute in Germany, as was the institute of experimental embryology established in 1889 (for Wilhelm Roux). Furthermore Breslau seems to have been unusually receptive to the introduction of applied sciences such as business economics (*Betriebswirtschaftslehre*) as well as positivistic traditions in the social sciences.[66]

But if one can judge from the Faculty's response to von Ruemker's reforms, this abundance of innovation was not necessarily due to 'progressive' attitudes within Breslau's Philosophical Faculty. On the contrary; he had to fight every inch of the way. As he confided to Friedrich Althoff, the senior civil servant in the Prussian Ministry of Education who was responsible for the universities,

> If I should conclude that I no longer have [the Ministry's] backing and support, my strength will not hold out much longer because the work and the struggles here have taken their toll, leaving me at times extremely tired, sleepless and overstressed. If I can count on further support, I can carry on a bit; otherwise I will soon be finished and will have to withdraw.[67]

65 *Mitteilungen [...] Breslau*, 1901; *Landwirt. Jahrbuecher*, 1903. By 1912 the process of specialisation had proceeded even further; even von Ruemker's own institute had four separate departments: for plant breeding, soil science (*Ackerbau*), agronomy, and chemistry, each with its own rooms, personnel and budgets (von Ruemker, 'Die Abteilung fuer Pflanzenzucht').

66 On physiology, chemistry and plant physiology, see Wolf, 'Die Koenigliche Universitaet Breslau' and Andreae, 'Schlesische Friedrichs-Wilhelm-Universitaet'. On embryology see Churchill, 'Wilhelm Roux'. On business economics see Locke, *End of Practical Man*, table 5.12. On positivistic social sciences see vom Bruch, 'Von der Sozialethik'.

67 'Wenn ich zur Ueberzeugung kommen sollte, dass ich [oben?] keinen Rueckhalt und keine Stuetze mehr habe, werden meine Kraefte nicht mehr lange

Although the reforms at Breslau reflect von Ruemker's vision, energy and skill at institution-building, he could never have accomplished them without Althoff's backing. As has often been noted, Althoff was an extremely powerful and skilled wheeler-dealer whose support for a variety of 'modernising' reforms played a major role in reshaping the Prussian university system from the 1880s to the First World War. To carry this off, he relied upon a network of trusted advisors from within the academic community who kept him informed about good people and new developments worth backing, and who in turn were guaranteed of Althoff's support for their own causes. Von Ruemker was one of those advisors, and he was very conscious just how much his organisational successes were dependent upon Althoff.[68]

With the creation of a chair for farm management in 1904, von Ruemker felt that the basic groundwork had been completed that would henceforth allow agriculture at Breslau to thrive. But his difficulties with the Philosophical Faculty were not over. That doctoral students had to pass an exam in philosophy was, he felt, 'a time-consuming ballast' and the amount of material to be mastered was excessive for a relatively short degree-course. Moreover, although he wished to make practical experience an admissions requirement, von Ruemker was unable to do so as long as agriculture was in the Philosophical Faculty. For its part, the Faculty was afraid that an increased number of agriculture students, lacking the full secondary diploma, was lowering its standing, and agricultural staff were given to understand that their field was something inferior which did not belong at a university, 'more a technical or commercial field than a scholarly'

vorhalten, denn ich bin durch die Arbeit und Kaempfe hier sehr mitgenommen, zeitweise aeusserst marode, schlaflos und ueberangestrengt. Kann ich vertrauen, dass [man?] mein Streben foerdert, so geht es noch eine Weile, sonst aber bin ich bald fertig, und kann oder muss mich ziemlich abgenutzt zurueckziehen' (14 Dec 1899, von Ruemker to Althoff, Althoff papers). Ruemker later recounted these struggles in his 'Organisation und Ziele'.

68 On Althoff see vom Brocke, 'Friedrich Althoff' and references therein. Von Ruemker's sense of indebtedness to his patron comes across most strongly in the aftermath of Althoff's success in pushing through an associate professorship in farm management for Aereboe: 29 Dec 1903 and 15 Sept 1904, von Ruemker to Althoff, Althoff papers.

one. Similarly, when seeking a successor for von Ruemker in 1913, the Faculty's appointments committee rejected one of the Ministry's nominees on the grounds that the candidate did not possess the *Abitur*. Matters were not helped by the fact that some in the Faculty also resented agriculture's abundant funding.[69]

By 1906 matters had escalated to the point that von Ruemker was appealing for Althoff's help in finding an alternative institutional home for agriculture:

> The Philosophical Faculty here is making such difficulties for us at every turn – they oppose the further development of agricultural sciences on grounds of the Faculty interest – that sooner or later our removal [from the Faculty] will be necessary [...]. Something will have to happen because relations within the Faculty are getting so tense that soon a way out will have to be found. Please be kind enough to give this question some thought and in due course to sort out the matter in a way which is most satisfactory for all parties.[70]

His preferred solution was that agriculture should be combined with economics (and perhaps forestry) to form a new 'Faculty of Economics'. Should that not be possible, Ruemker reckoned, one could explore the possibility of an agricultural faculty at the new engineering college planned for Breslau. The least attractive option was to establish an agricultural college in Breslau.[71]

69 Ruemker, 'Organisation und Ziele'. 'Time-consuming ballast' is from 17 Dec 1905, von Ruemker to Althoff, Althoff papers. On von Ruemker's successor, see 25 Feb 1913, Philosophische Fakultaet to Kultusministerium, GStaA, Rep 76, Va, Sekt 4, Tit IV, Nr. 48, Bd. II. 'More a technical or commercial field' is from Kaufmann, *Festschrift*, 233.

70 'Die hiesige philosophische Fakultaet macht uns derartige Schwierigkeiten bei jeder Gelegenheit weil sie aus Fakultaetsruecksichten eine Weiterentwicklung der Landwirtschaftswissenschaft nicht wuenscht, dass ueber Kurz oder Lang die Herausloesung zur Notwendigkeit werden wird [...]. Indessen geschehen wird etwas muessen, da die Verhaeltnisse in der Philosophische Fakultaet sich derartig zuspitzen, dass in Balde ein Ausweg gesucht werden muss. Vielleicht haben Eure Exzellenz die Guete, die Frage im Auge zu fassen und zu gegebener Zeit die Gestaltung so durchzufuehren, wie sie fuer alle beteiligten Kreise die wuenschenswerteste und nuetzlichste ist.' (von Ruemker to Althoff, 4 Mar 1906, Althoff papers).

71 4 Mar 1906, von Ruemker to Althoff, Althoff papers. According to Aereboe, von Ruemker had also made headway at some point by demanding a separate

At this point let us pause briefly and consider these local events in the context of the national agenda for agricultural education. For although conflict between the agricultural institute and the Philosophical Faculty was not unique to Breslau, it is quite likely that these events prompted von Ruemker's published proposals for the reform of agricultural higher education which we encountered in Chapter Three. As we saw there, for example, it was in 1906 that he published his proposal for the establishment of 'faculties of economics'. It was essential, he wrote at that time, that agricultural science could 'develop unhindered in all of its branches' and that the curriculum contained 'only what is really essential and practically valuable [...]. [Subjects should be avoided] which are of no use to the agriculturist and are only taken "for the sake of form"'.[72] The prospects of this happening at Breslau, however, cannot have looked particularly rosy, for a few years earlier Breslau's senate had rejected a proposal to establish a 'faculty of technology' on the grounds that this was not in keeping with the nature of a university. Indeed, this is why plans were being made around 1906 to create a new engineering college in the city, though it turned out to be a very modest institution with only two faculties. And in 1907 – just a year after von Ruemker's proposals had appeared – Althoff retired, depriving Ruemker of a crucial ally. Against this backdrop, one can understand why by 1909 he had changed his mind. New faculties in either the universities or the engineering colleges, he now wrote, were unlikely. The best solution was to combine agriculture with forestry and veterinary medicine into large independent colleges (*Landeskulturhochschulen*), one aim of which would be to free the agricultural sciences from their 'continuing powerlessness'.[73]

'Faculty of Agriculture' (Aereboe, 'Landwirt. Hochschulen', 642–643). Interestingly, the proposed 'Faculty of Economics' found support from Julius Wolf, professor of economics, whose positivist sympathies were at that time quite far from the mainstream of German economics (von Ruemker, 'Organisation und Ziele'; vom Bruch, 'Von der Sozialethik').

72 Von Ruemker, 'Zur Frage der Ausgestaltung', 89.

73 Von Ruemker, 'Ueber die Entwicklung'; see also his 'Landeskulturhochschulen?'. On the new engineering college see Manegold, *Universitaet*, 195. The institution which at that time most closely approximated to Ruemker's

However tense the agricultural institutes' relations with the rest of the Philosophical Faculty were, their ties to the farming community were excellent. Next door to the institutes was the headquarters of the Silesian Chamber of Agriculture with its agricultural experiment station and an important seed control station. Connections between the institutes and the Chamber were close from the start, not least because until 1920 the institutes were housed in two buildings belonging to the Chamber. From 1895 the institutes contributed annually to short courses for farmers organised by the Chamber.[74] In addition to organising a number of short courses for plant-breeders, von Ruemker claimed to have given over two hundred lectures to agricultural organisations between 1895 and 1910.[75] Many of the Chamber's scientific staff had studied agriculture at the University, and some shared research interests with their academic neighbours. The Chamber awarded both student fellowships and annual prizes for the best students' work which advanced Silesian agriculture, and it made its experiment stations available for teaching purposes. On the occasion of the University's 100th anniversary in 1911, the Chamber donated a sizeable sum to be used for new agricultural buildings. Also nearby was the Agricultural Association of Breslau (*Breslauer Landwirt. Verein*) at whose monthly lecture-evenings agricultural professors often spoke and whose support may have played a role in the creation of a chair for farm management in 1904.[76]

One reason for these cordial relations may have been the practically oriented character of Breslau's curriculum. As we saw in Chapter

vision was the Agricultural College in Berlin with its three faculties and a neighbouring Veterinary College. That is perhaps one reason why in 1912 he accepted a call to the chair of agronomy there.

74 On housing see Zorn, 'Die neuen landwirt. Institute'; on the Chamber's short courses see *Landwirt. Jahrbuecher,* 1903.

75 On plant-breeding courses, see *Landwirtschaftliche Jahrbuecher*, 1910 and Oldenburg, *Das landwirt. Unterrichtswesen*; on lectures see von Ruemker, 'Breslau', 45.

76 Zorn, *Geschichte der Landwirtschaftswissenschaften*, 93–97; Holdefleiss, 'Das landwirt. Studium'. On farm management see 29 Dec 1902, Vorstand des Landwirt. Vereins in Breslau to Kultusministerium, GStaA, Rep 76Va, Sekt 4, Tit IV, Nr 36, Bd xxi.

Three, von Ruemker placed considerable emphasis upon farm management as a foundational subject, and Breslau was the first of the universities to acquire a separate chair for the subject. He also repeatedly stressed the importance of students obtaining practical experience in agriculture prior to their university studies. A 'critical understanding' of lectures was not possible, he insisted, unless one had worked in agriculture for at least three to four years. And while he rejected the idea that the farming community should prescribe the issues which academics were to teach and research, he believed it was an academic's duty to maintain close ties with farmers and to address their problems rather than 'retreating into the quiet isolation of the scholar's study'. These views were evidently appreciated by the Chamber of Agriculture who petitioned the Ministry to retain von Ruemker in 1905 when he was offered a chair at Bonn.[77]

Nor was von Ruemker the only member of Breslau's staff with a strong practical orientation. It is no accident, for example, that its first (associate) professor of farm management was none other than Friedrich Aereboe (cf. Chapter Two) whom von Ruemker had discovered while the former was working as director of a large estate in Silesia. Though Aereboe only stayed for two years, von Ruemker was impressed by his book on the principles of farm management, and Aereboe was brought back from 1913 to 1919 as full professor. During the interim period 1906–1913 Breslau appointed Franz Waterstradt who had been working at the German Agricultural Society. A book which he published in 1912 has been regarded by some historians as the first textbook of modern farm management.[78] Thus Breslau, along with Bonn, was a centre for innovative work in this practice-oriented field before the First World War.

77 'Critical understanding' is from von Ruemker, 'Programm fuer das Studium', 15; see also his 'Organisation und Ziele'. 'Quiet isolation' is from von Ruemker, 'Das landwirt. Versuchs- und Bildungswesen', 104; cp 109–110. On the Chamber's support for von Ruemker see 17 Jan 1905, Landwirtschafts-Kammer to Min. Landwirtschaft, GStaA, Rep 76, Va, Sekt 4, Tit IV, Nr. 23.

78 On Aereboe see Ries, 'Zum hundertsten Geburtstag'. On Waterstradt see Zorn, *Geschichte der Landwirtschaftswissenschaften* and Frauendorfer, *Ideengeschichte*, 475.

That Waterstradt was appointed after Ruemker's departure for Berlin indicates that Breslau's practical orientation was not due solely to von Ruemker's influence. When Ruemker left, the Faculty initially tried to hire Erich Tschermak. As I have argued elsewhere, although Tschermak is generally credited as one of the rediscoverers of Mendelism, his approach to plant-breeding was shaped mainly by practical concerns.[79] On receiving his doctorate in plant physiology in 1896, for example, his aim was not to remain in academe but to become a horticultural breeder. And when he stumbled upon Mendel's work a few years later, he saw it as significant, less for genetic theory than as a labour-saving device for breeders. As a breeder, moreover, he enjoyed considerable commercial success, several of his cereal varieties being planted on a large scale into the 1960s. In order to be an effective plant-breeder, as Tschermak saw it, a knowledge of Mendelian genetics was necessary but by no means sufficient; other kinds of skill were equally essential, and breeding was an art rather than a science. In the event, Tschermak chose to remain in Vienna, so Breslau appointed Friedrich Berkner (1874–1954). The son of a small farmer, Berkner had spent his previous career in school teaching and in a lower level agricultural school in East Prussia where he was also active as an advisor on agricultural policy. Although aged 39 at the time of his consideration for the chair at Breslau, Berkner had very few scholarly publications to his name, a fact which initially aroused opposition within the Faculty to his being shortlisted. Agricultural staff on the appointments committee, however, stressed the importance of his practical experience, and the Minister was eventually persuaded to appoint him.[80]

79 Harwood, 'Rediscovery of Mendelism'. On Tschermak's dealings with Breslau see Frimmel, 'Hofrat'.

80 On Berkner's career see Boehm, *Biographisches Handbuch* and Zorn, *Geschichte der Landwirtschaftswissenschaften.* On his appointment at Breslau, see 25 Feb 1913, Pfeiffer to Kultusminister, GStaA, Rep 76, Va, Sekt 4, Tit IV, Nr 48, Bd II. (For similar concerns at an earlier stage in the appointment, see 10 Aug 1912, Philosophische Fakultaet to Kultusminister, in same file.) Significantly, when Berkner was being considered for chairs elsewhere after the First World War, one of his strongest supporters was that apostle of 'practice', Friedrich Aereboe: see PrA Phil Nr. 15 and Phil k21 (Personalakte Hermann

Thus agriculture at Breslau developed in a practically-oriented way *despite* continued opposition from its Philosophical Faculty, thanks partly to a combination of von Ruemker's determination and political skill plus Althoff's sympathetic support. The nature of the regional economy seems not to have been particularly important, for in this respect Breslau was quite similar to Halle. The proportion of large farms in Silesia was relatively high (cf Table 2.1), and agricultural industry was well-represented there. Bordering on a sugar-beet-growing area, Breslau had many sugar-factories, and there were several major plant-breeding firms nearby. Moreover the city hosted annual markets for agricultural machines and for cattle.[81] What does seem to have distinguished Breslau, however, was the character of its University. For unlike cosmopolitan universities like Berlin, Halle or Heidelberg, Breslau was very much a locally oriented institution which enjoyed substantially less academic prestige. Throughout the 19th and well into the 20th century the great majority of its students (including its agriculture students) came from Silesia (eg, over 70% in 1911 and in 1928), and the University played a major role in supplying the region with its professionals and civil servants. By contrast, only about 30% of Halle's agriculture students came from the province where it was situated. Conversely, while typically over 30% of Halle's agriculture students came from outside Germany, only 2–3% of Breslau's did.[82] When von Ruemker arrived in Breslau, therefore, this tradition of servicing the regional economy – augmented, no doubt, by the presence in the city of major agricultural organisations which took an active interest in teaching and research there – meant that von Ruemker and his colleagues had to reckon with relatively strong pressure to address local concerns. But the Institute which he

Kraemer), both UA-Giessen. Berkner's marked interest in economic questions is thought to have made him an attractive candidate before and after the First World War (Meyer, 'Prof. Dr. F. Berkner').

81 See Mitteilungen [...]Breslau, vol 2, 1902.

82 On the University's local character, see Andreae, 'Schlesische Friedrich-Wilhelms-Universitaet' and Zorn, *Geschichte der Landwirtschaftswissenschaften*, 84. On students' regional origins see *Deutsche Hochschulstatistik* for summer semester 1928. On non-German students for 1900–1908, see *Landwirt. Jahrbuecher*, 1910.

took over also lacked Halle's international reputation as well as its academic prestige. Rather than trying to compete with Halle on those terms, therefore, von Ruemker set out to exceed Halle in terms of size and practical facilities while relying on Breslau's university status to differentiate his institutes from the academies. To judge from contemporary accounts, he succeeded.

The Agricultural College at Hohenheim

A tour of the major German agricultural institutions before the First World War would not be complete without Hohenheim. On the one hand Hohenheim was one of those 'isolated academies' which Liebig had damned, however unfairly, in the 1860s, and it never enjoyed the academic prestige of the colleges at Berlin or Bonn. But in fact around the turn of the century Hohenheim enjoyed some of the best teaching and research facilities in Germany, plus well-respected staff, so that it was able to attract about one-third of its students from outside Germany.[83] To be sure, those adhering to a 'scientific' vision of agricultural education were not impressed. Thus when C.B. Hutchison and Albert Mann visited the campus in 1927 on behalf of the International Education Board, they concluded that Hohenheim (as well as Weihenstephan) 'can hardly be called more than trade schools of a higher type. They are training skilled farmers rather more than educating men and women broadly for rural life.'[84] But it is precisely Hohenheim's

83 As we will see in Chapter Five, when a Bavarian committee charged with the restructuring of agricultural education during the 1920s wanted to inform themselves about the requirements of a well-equipped institution, the places they chose to visit were Halle and Hohenheim.

84 Hohenheim won praise in 1895, for example, from Franz Soxhlet, professor of agriculture at the engineering college in Munich ('Aeusserungen des Prof. Dr. Soxhlet, betreffend die landwirt. Zentralschule in Weihenstephan […] 13. Juli 1895', TUM-Archiv, Sign = I, 8, Bd 1). On student nationality see *Jahresberichte-Hohenheim*, 1900 to 1914. The quote is from Hutchison and Mann, 'Higher education in agriculture', 51.

combination of low status and a strong practical orientation which makes it an important case study.

Hohenheim was established as an agricultural school in 1818, modelled upon Thaer's academy at Moeglin. During the 1840s and 1850s several of Hohenheim's staff (H.W. Pabst, the agricultural chemist Emil Wolff, and the director Gustav Walz) had conducted extensive experiments on the efficacy of Liebig's mineral fertilisers and had found them wanting. And it is probably for that reason that Liebig's sharpest criticism of the academies during the 1860s was directed at Hohenheim. In the aftermath of that clash, however, Hohenheim did not suffer the fate of most of the other academies but was strengthened in several respects. Besides acquiring an additional chair of agriculture, provision for the natural sciences was improved by establishing an experiment station for agricultural chemistry (1864) as well as chairs for zoology and mineralogy (ca 1870). In addition the Wuerttemberg state government chose to enhance Hohenheim's academic status in a number of ways: by transferring it from the Ministry of Agriculture (*Landwirt. Zentralstelle*) to that for Education, by conferring upon it the name and constitution of an 'academy', and by granting its professors the rank of those at Wuerttemberg's university in Tuebingen.[85]

Nevertheless, Hohenheim still did not enjoy the academic status of its sister colleges at Berlin or Bonn. For one thing, it was too far away from Tuebingen (and a bit too far from Stuttgart's engineering college) for its students to be able conveniently to attend non-agricultural courses there, so that it was not in a position to offer an 'all-round education', as critics were fond of pointing out. And unlike Berlin and Bonn, Hohenheim offered only the four-semester degree (with its lower entry requirement) until shortly before the War. Furthermore in terms of formal academic status, it was not upgraded from an academy to a college until 1904, some twenty years after Berlin, nor did its students have the opportunity to take a doctorate (at

85 Franz, Liebig und Hohenheim; Kirchner, Die Entwicklung der Koeniglichen.

Tuebingen) until 1914, over a decade after students at Bonn or Berlin.[86]

Needless to say, this status-deficit made it difficult for Hohenheim to recruit staff with high academic qualifications (which reduced its status yet further). Very few of its staff in the agricultural sciences had come from universities or engineering colleges (and the few who did had left a junior position in order to take up a chair at Hohenheim). The great majority had previously been employed in agricultural administration, schools, estate-management or experiment stations. In addition over one-third of Hohenheim's staff lacked doctorates at the time of their appointment, and few had habilitated.[87] On the other hand, staff in economics or the basic sciences generally had stronger academic qualifications, but attracting them to Hohenheim was not always easy. In the late 1880s, for example, the College spent two years trying to find a highly qualified person to teach mathematics and physics and finally ended up hiring a school science teacher. And when the chair of economics fell vacant in 1903, it took three years to find a suitable candidate who then accepted the post but left for a university chair before even turning up in Hohenheim. When the College did succeed in recruiting well qualified academics, about a quarter of them moved on within five years to universities or engineering colleges.[88]

86 Among the critics was von Ruemker, *Die Ausbildung des praktischen Landwirts*. On the four-semester degree see Springer, 'Die Koeniglich Wuerttembergische' and Kirchner, *Die Entwicklung der Koeniglichen*. On upgrading see Klein, *Die akademischen Lehrer*.

87 This generalisation is based on professorial staff in the agricultural sciences (omitting those in forestry and veterinary medicine) who were in post between 1890 and 1915, using data from Klein, *Die akademischen Lehrer* and Franz, 'Geschichte der Universitaet'. The same picture emerges from a survey of the *Jahresberichte-Hohenheim*, 1900–1914. A case in point was the agronomist and plant-breeder, Carl Fruwirth (cf. section 6 below). When appointed in 1897 he had no doctorate and acquired one only ten years later after petitioning the Education Ministry (Dolezal, 'Hofrat Prof. Dr.').

88 This is based on career data for economists and basic scientists who were in post between 1890 and 1915, published in Klein, *Die akademischen Lehrer*. On appointments in physics and economics see ibid., 21, 24–25, 28.

Though lacking in academic status, Hohenheim was strong in other respects. By the eve of the First World War the College was just as large and well-equipped as Bonn, with a large model farm (300 ha) plus experimental grounds for agronomy (8 ha) and agricultural chemistry (2 ha) and a botanical garden (5 ha). With twelve professors, it was second only to Berlin in the size of its staff. Moreover Hohenheim sought to address the needs of the farming community in various ways. Like most other institutions, the College offered short courses for working farmers (from 1891), but more significant is the fact that a very high proportion (about three-quarters) of the articles published by staff appeared in outlets accessible to the farming community.[89]

The economic dimension of agriculture was also taken seriously. When the professor of economics (*Nationaloekonomie*) retired in 1903, the Ministry considered eliminating the post, but staff protested that economics had by then become more important than ever and that all urgent agricultural issues were economic rather than technical in nature. Accordingly, farm management also maintained a high profile at Hohenheim. In 1884 the College was one of the first institutions to dedicate a separate chair to the subject, and thenceforth the chair was routinely held by the College's director. In 1919 Friedrich Aereboe, at that time Germany's most eminent figure in the discipline and a vigourous defender of practically-oriented education, was appointed as Hohenheim's Director.[90]

The structure of Hohenheim's curriculum reflected this practical orientation. Unlike Halle, the College did not discourage students from taking agricultural courses in their first year. In 1889, for example, students in the first semester of the two-year course encountered not only natural sciences, but also plant production, animal

89 This estimate is based on information published in Jahresberichte-Hohenheim, 1900 to 1914. The accessible outlets included Wuerttembergisches Wochenblatt, Deutsche Landwirtschaftliche Presse, Allgemeine Fischerei=Zeitung, and Allgemeine Molkerei=Zeitung. On facilities and staff numbers see Mentzel & Lengerkes Schreibkalender, 1913; and Oldenburg, Das landwirt. Unterrichtswesen. On short-courses see Kirchner, Die Entwicklung der Koeniglichen.

90 On economics see Franz, 'Geschichte der Universitaet', 93. On farm management see Boelcke, 'Die Gruendungsgeschichte'; and Klein, *Die akademischen Lehrer*, 23.

husbandry and agricultural machines. In 1913 students on the three-year course were taking cultivation technology and horticulture in the first semester.[91] When it came to science-teaching, Hohenheim – unlike the universities or Berlin – was keen on tailoring such teaching to agricultural problems. Indeed the College used this as a selling point in their brochures aimed at students. Whereas Julius Kuehn at Halle referred to the basic sciences in the context of the curriculum as 'fundamental sciences' (*Grundlagenwissenschaften*), at Hohenheim they were described as 'ancillary sciences' (*Hilfswissenschaften*), thus reflecting the differing visions of agricultural education at the two institutions. With this aim in view, when appointing science staff Hohenheim looked for candidates with applied (and if possible agricultural) interests who would be able to maintain good relations with the farming community.[92]

One reason why Hohenheim maintained such close ties to practice was the fact that many of its institutes were formally linked to state experiment stations (*Landesanstalten*) so that the institute's professor also served as director of the station. Thus Wuerttemberg's plant-breeding station (est. 1905) was an integral part of the College's institute of agronomy, just as the state's seed testing station (est. 1878) was part of the institute of botany.[93] These stations were required to serve the state's residents and did so by carrying out tests on feed, fertiliser, milk, seed, machinery and so forth. The stations were also

91 Uebersicht ueber die Organisation, 1889; Anon., Das Studium der Landwirtschaft .

92 On basic sciences at the two institutions compare Anon., *Das Studium der Landwirtschaft*, 6, with Kuehn, *Programm fuer das landwirt. Studium*, 16. On appointments criteria see 21 June 1909, Kultusminister to Direktion der Landwirtschaftlichen Hochschule, UA-Hohen, Personalakte-Hermann Kraemer. On botany appointments see 22 Jan 1917, Direktor of Hohenheim to Kultusminister, UA-Hohen, AZ=4/1 (Personalakte Georg Tischler). Although Kirchner (*Die Entwicklung der Koeniglichen*) claims that science teaching was not tailored in 1860s since it was also taught to forestry students, this may not have lasted long since forestry was removed from Hohenheim in 1881.

93 Steiner, 'Zur Geschichte der Saatgutpruefung'; Wacker, 'Die Kgl. Wuerttembergische Saatzuchtanstalt'. Additional stations for machinery-testing (1883), the fermentation industry (1889), and the dairy industry (1892) were also established (Franz, 'Geschichte der Universitaet', 84, 88).

quite important for Hohenheim's material wellbeing since the fee income from testing was considerable, contributing over half of the College's income by the eve of the First World War. Similarly, it was through the establishment of the plant-breeding station, for example, that the Institute of Agronomy got its first purpose-built quarters.[94]

Unlike Breslau, however, the origins of Hohenheim's students were probably not very important in keeping staff attuned to local agricultural issues. For in the decades before the First World War only about 30% of the College's students came from Wuerttemberg while nearly as many (25%) came from outside Germany. On the other hand, most of the *staff* were from Wuerttemberg during the period 1890–1914, apparently because of political pressure.[95] Since agriculture in Wuerttemberg was overwhelmingly dominated by small farms (cf. Table 2.1), it is perhaps significant that some of the staff were the sons of peasants. The director and professor of farm management (1915–1919), Hermann Warmbold, was a case in point.[96] At Hohenheim he set up an advisory service for small farmers and developed a system for the efficient use of grassland (using it alternatively for hay and as pasture) which acquired international recognition.

But even staff from other social backgrounds addressed their research to regional problems. Franz Waterstradt (professor of farm management 1913–1914), for example, was particularly interested in the problems of small farmers. The same was true for Adolf Muenzinger (professor of farm management from 1922). Before coming to Hohenheim Muenzinger had been employed for over a decade as manager of several large estates. Regarding himself as a practical man, he was not sure in 1922 that he wanted to move to academe but

94 On fee income see Kirchner, *Die Entwicklung der Koeniglichen.* On agronomy see 'Die Entwicklung des Instituts fuer Acker- und Pflanzenbau', typescript, n.d. but 1960s, UA-Hohen, Sign= 3/25.31.

95 On students' origins see Kirchner, *Die Entwicklung der Koeniglichen.* On staff origins see Franz, 'Geschichte der Universitaet', 87, 90, and Behrend von Grass, 'Erinnerungen'. During the latter half of the 19th century it is said that only some 40% of staff were Wuerttemberger, though nearly all directors were (Klein, *Die akademischen Lehrer*, 25).

96 Klein, Die akademischen Lehrer, 27, and Haushofer, Furche der DLG, 190.

was attracted to Hohenheim because of what he saw as the close ties there to the farming community. Feeling that farm management had focussed heretofore largely upon large farms, Muenzinger and his students began to study the problems of small family-run farms in order to find out how such farms might take advantage of intensive cultivation methods. One practical outcome was a cooperative (for sharing machinery) which he set up in the early 1930s with government assistance and which is said to have attracted international attention. These efforts on behalf of small farmers attracted favourable attention in Bavaria, another state whose agriculture was dominated by small farms (cf. Chapter Five).[97]

In other cases staff had a great deal of experience in practical agriculture before coming to Hohenheim. Otto Vossler, for example, had worked in farming for ten years before being appointed director in 1884. In addition to seven years in farming, Hermann Warmbold had spent eight in agricultural administration prior to his arrival at Hohenheim in 1915. On leaving four years later, he became a top civil servant in the Prussian Ministry of Agriculture. Similarly, when in 1907 the College was appointing someone to the chair of agronomy, none of the shortlisted candidates then occupied an academic post. One (Hans Lang) was director of plant-breeding at a large estate; another (Fritz Franck) worked for the Chamber of Agriculture in Saxony; and the third (Johann Wacker) had been a teacher, estate-manager and agricultural inspector for the last fifteen years. Moreover all three were from Wuerttemberg because the appointments committee was agreed that the professor of agronomy ought to be familiar with local agricultural conditions.[98]

97 On Muenzinger see his autobiography, *Aus meinem Leben;* Boehm, *Biographisches Handbuch*; and Klein, *Die akademischen Lehrer*, 30. On Bavarian interest in Hohenheim see 14 May 1931, Paul Kulisch to Kultusminister, BayHStA, MK 19684.

98 On Vossler see Franz, 'Geschichte der Universitaet'. On Lang and Franck see the memoirs of Oskar von Kirchner, pp 610–614, UA-Hohen, AZ=N5/1.2 (Kirchner papers). On Wacker see *Kuerschners*, 1931. On the appointment more generally see AZ = 4/1 (Johann Wacker papers), UA-Hohen. Interestingly, one reason why Erich von Tschermak was not shortlisted for this job is that he was not a Wuerttemberger; another is that several members of the

At Hohenheim, therefore, the College's geographic location made a 'practical' orientation possible; the social origins of staff and their extensive experience outside academe enabled this possibility to be realised; and the dominance of peasant-farming made it difficult to avoid. But of course staff with such backgrounds did not gravitate toward Hohenheim spontaneously; someone had to appoint them. And it is clear that the farming community, aided by the Education Ministry, was able to exert significant influence upon the College in a number of instances, sometimes overriding resistance from staff themselves. In 1881, for example, the Ministry chose to appoint Ernst Strebel – who had worked over the previous decade as an estate manager and as a school teacher – as professor of agriculture. As concerned with academic status as those elsewhere, Hohenheim's staff objected that Strebel had been educated at an agricultural secondary school (rather than at a grammar school), had no doctorate, and had published very little, but the Ministry simply ignored them. Fifteen years later when a new director was to be appointed, the majority of staff again wanted a man with a 'scientific reputation'; few felt that the appointee needed to have had experience of running a farm. But the Ministry was keen that the new director should be familiar with conditions in Wuerttemberg, and it was again Strebel (a Wuerttemberger) who got the job.[99]

Another arena in which the influence of regional interest-groups was much in evidence concerned the College's governance. At Hohenheim staff had begun to press for a rectorial constitution in 1882 (Berlin had been granted one the year before), but farmers feared that the change would weaken the College's ties to agriculture, and thus the staff's request was rejected by both houses of Wuerttemberg's legislature. (The legislature's finance committee argued that the proposal was inappropriate because Hohenheim was 'not a real college' (*Hochschule*) and suspected that it was anyway prompted merely by

committee were also concerned that he was more a botanist than an agricultural scientist.

99 Franz, 'Geschichte der Universitaet'; Klein, *Die akademischen Lehrer*, 23–24.

staff concerns for academic prestige.)[100] The issue reemerged in 1910 with unanimous staff backing and was once again rejected by the Minister. Since the head of the College had to have the respect of Wuerttemberg's farmers as well as practical experience sufficient to oversee the College's model farm, the Minister argued, the power to appoint him must remain with the Ministry. Two years later the item was back on the agenda and was once again rejected because farmers had reservations about 'strengthening the purely academic element' at Hohenheim. After the First World War Friedrich Aereboe was appointed director and was immediately confronted with a staff campaign for a rectorial constitution. Although farmers' organisations still preferred the directorial constitution, by this time the Association of German Higher Educational Institutions (*Verband der deutschen Hochschulen*) was refusing to admit any institution to membership unless it had a rector. Faced with this situation, the Ministry finally caved in, arguing that by then all other agricultural colleges had acquired a rectorial constitution and that Hohenheim should not appear to be a second class institution.[101]

The tension between staff concerns for status versus accountability to political and economic constituencies was also evident in a further clash between Aereboe and his staff: namely, over his request that an agricultural advisory board be created. Aereboe had made it a condition of his accepting the directorship in 1919 that a board representing farmers from Wuerttemberg and Baden would be set up to advise the Ministry on educational matters. As we have seen, such a body was resisted at Berlin, but at Hohenheim (as at Bonn) one was established in 1920. Although the board included four staff from the College, they were outnumbered by four farmers, an industrialist, a representative from the Ministry, and four representatives from the Chambers of Agriculture in Wuerttemberg and in Baden.[102] This body

100 26 Sept 1913, Kultusminister to King, Wuertt. HStaA, Bestand = E14, Sign = Bue 1614.

101 18 May 1922, Hieber to Kultusminister, Wuertt. HStaA, Bestand = E 130b, Sign = Bue 1519. See also Franz, 'Geschichte der Universitaet', 97–100; Klein, *Die akademischen Lehrer*, 26; Aereboe, *Jugendlust*, 152.

102 Satzung dated 21 Dec 1920; 29 Dec 1920, Kultusminister to Director of College, both in UA-Hohen, AZ = 3/02.12.02.

had substantial powers; if its recommendations were endorsed by the Minister and the director, they could be implemented despite objections from academic staff. At its first meeting in 1921 the board duly exercised these powers. Addressing the question of College entry requirements, the board concluded that although requiring the advanced secondary school diploma (*Abitur*) was a good idea in principle, it would effectively keep out able farmers, and the deciding factor had to be the needs of agriculture rather than the academic status of students and the College.[103] Following Aereboe's departure in 1922 and the election of Hohenheim's first rector, however, the existence of the advisory board was challenged by academic staff . The College's relations to the region, they argued, ought to be primarily of a scholarly (*geistiger*) nature. Hohenheim should be a leader of agriculture and the agricultural sciences rather than a mere 'dispenser of mouse poison'. But their challenge failed; Wuerttemberg's farmers continued to regard the board as useful, and it was retained. In 1931 once again several staff tried to get the board abolished on the grounds that it was 'unworthy of a college', but again the Minister declined to oblige them.[104]

Although many staff at Hohenheim were genuinely concerned to serve Wuerttemberg's farming community, these episodes show that – like their colleagues in Berlin – they were *also* sensitive to matters of academic status and sought accordingly to enlarge the College's autonomy. But their efforts were curtailed through the combined action of regional interest-groups plus ministerial resolve. With agriculture dominated by small farms, farmers' organisations took an active interest in Hohenheim, blocking changes which they felt would buffer it from their concerns. In addition, evidently determined to ensure that Hohenheim did not lose touch with the world of practice, Wuerttemberg's Education Ministry was prepared to site a number of experiment stations there and to resist (at least for a while) staff

103 Minutes of meeting on 25 Jan 1921, UA-Hohen, AZ = 3/02.12.02. On the advisory board's powers see Franz, 'Geschichte der Universitaet', 105.

104 'Besprechung' dated 26 Apr 1922; minutes of meeting of 11 July 1923; and minutes of meetings of Senat on 6 May 1931 and 24 June 1931; all in UA-Hohen, AZ = 3/02.12.02.

attempts to shift the College in a more 'science-oriented' direction. Despite formal upgrading, therefore, the cognitive aspects of academisation could be partly contained.

Plant-breeding research at the five exemplary institutions

In characterising the science/practice orientation of these institutions, we have focussed primarily upon such features as curriculum, staff backgrounds and demonstration facilities, though there were indications at some places that research was also skewed in the same direction. I now want to look more closely at research activity at these institutions, though a detailed study of all of the agricultural disciplines is obviously impossible. The major agricultural disciplines are simply too diverse for any one historian to manage. I will focus, therefore, just upon plant-breeding since it lends itself nicely to this task. After the rediscovery of Mendel's work in 1900, academic plant-breeders the world over began to explore the implications which this new body of theory might have for breeding practice, and interestingly, their assessments of Mendelism's importance varied widely. This was evident, for example, in their views as to whether successful breeding was an 'art' or a 'science'. In Germany Erwin Baur insisted that 'every advance in plant- and animal-breeding rests exclusively upon experimental genetic research.'[105] Erich Tschermak, on the other hand, took the view that breeding, 'is an art which requires a particular gift and a "breeder's eye"'.[106]

105 Baur, [opening address], 52.
106 Tschermak, 'Ein Leben fuer die Zuechtung', 777.

Table 4.1 – Academic Breeders Take Sides

Enthusiasts	*Sceptics*
1. How important was hybridisation?	
Baur	von Ruemker
Roemer	Fruwirth
Kiessling	Wacker
Tschermak	Edler
Sessous	Kulisch
	Wohltmann
	Kraus
	Schindler
	Lang
	Froelich
2. Was selection in pure lines powerless?	
Baur	Fruwirth
Roemer	von Ruemker
Riede	Hillmann
Kiessling	Sessous
	Wacker
	Kraus
	Wohltmann
	Edler
	Tschermak
3. Hybred varieties or landraces?	
Baur	Fruwirth
Roemer	Wacker
	von Ruemker
	Kiessling
	Kraus
	Tschermak
	Schindler
	Lang
	Liebscher
	Kulisch
	Jelinek
	Gisevius
	vonLiebenberg
	Ziegler
	Kryzymowski

These differences of judgment as to the value of Mendelism are also noticeable in several debates which divided the German academic plant-breeding community from the turn of the century into the 1920s (Table 4.1).[107]

The first of these centred on the value of hybridisation (generally perceived to be the'Mendelian' method) versus pre-Mendelian methods such as individual selection or mass selection. The second was triggered by the Danish geneticist, Wilhelm Johannsen's, 1903 argument that selection in pure-lines was powerless (a claim which was especially controversial in Germany since it contradicted the common practice among late 19th century commercial breeders of *repeated* selection, even in allegedly pure lines). And the third revolved around the kinds of varieties thought most appropriate for boosting Germany's agricultural productivity. Some endorsed the 'hybred' (ie, highly bred) varieties developed by commercial breeders on the expectation that they would be superior in almost all growing conditions while others favoured locally adapted traditional varieties ('landraces') as the starting point for varietal improvement. Each of these debates pitted researchers who were confident of the utility of genetic theory (I will call them 'Mendelian enthusiasts') against those who were doubtful ('Mendelian sceptics').

As we shall see, these figures' views on the role of scientific theory in plant-breeding fitted closely with their more general conceptions of agricultural science and education. Moreover, especially important for the argument of this chapter is the fact that the most prominent enthusiasts among them worked at 'scientifically-oriented' institutions while many of the sceptics were employed at practically-oriented ones.

Probably the most outspoken of the Mendelian enthusiasts was Erwin Baur, professor of genetics at the College of Agriculture in Berlin from 1914, then director of the newly established Kaiser-Wilhelm Institute for Breeding Research (*Zuechtungsforschung*) from 1928 until his death in 1933. It is not surprising that he, perhaps more than any other academic breeder of his time, should have emphasised the importance of genetic theory. For Baur was also a leading figure in

107 For a fuller discussion see Harwood, 'The reception of genetic theory'.

the German genetics community, by virtue of his research accomplishments as well as his skill as an institution-builder.[108] While Baur was by no means indifferent to the needs of practice,[109] the practical concerns which he addressed were not those of the average farmer but of the commercial breeder and estate-owner (and this sector of the agricultural community, as we have seen, was particularly receptive to the value of science). From the start, for example, he maintained close ties to the main association of commercial breeders (*Gesellschaft zur Foerderung Deutscher Pflanzenzucht*), encouraged perhaps by the fact that it was intensive lobbying by this association which had led to the creation of his chair at the College in 1914. He also undertook contract-research in plant-breeding for a large estate-owner in Brandenburg after the First World War.[110]

The other major Mendelian enthusiast, Theodor Roemer, was professor of agronomy and plant-breeding at Halle from 1919. Like Baur, Roemer had close ties to the private-sector, though not to the farming community. Prior to his appointment at Halle, he had been head of breeding at a well-known commercial breeding firm. At Halle his attempts to develop laboratory methods for establishing a soil's fertiliser requirements soon attracted the attention of an association of fertiliser manufacturers (*Deutscher Kali-Syndikat*) who offered him a position as director of its agricultural division. Though he declined, Roemer served as a consultant for the association for several years. He also enjoyed exceedingly good ties to commercial breeders. Nearly all of the new cereal varieties which his institute produced during the 1920s and 1930s were sold off (in rough form) to private breeders for finishing off and multiplication.[111] To classify Roemer, along with Baur as a 'scientifically' oriented figure, therefore, is not to suggest that he was tucked away in an ivory tower far from the madding crowd. On the contrary; both men were intensely engaged with the world of 'practice', but for both of them this meant commercial

108 On Baur see Harwood, *Styles of Scientific Thought*, 229–245.

109 Eg, Baur, 'Zukunfts-Aufgaben'; Baur, *Die wissenschaftlichen Grundlagen*, preface and 84.

110 For details see Harwood, 'Politische Oekonomie'.

111 Harwood, 'Politische Oekonomie'.

breeders, agricultural industry or estate-owners – all of whom were keen to reap the benefits of scientific agriculture – rather than the vast majority of farmers who, as small land-holders, were not in a position to do so.

If Baur and Roemer were outspoken enthusiasts for Mendelism, others such as Ludwig Kiessling were more qualified in their support.[112] Kiessling was employed at the Bavarian State Plant-Breeding Station from 1902, becoming director and a professor at Weihenstephan in 1910. In a number of ways Kiessling self-consciously sought to straddle the divide between 'science' and 'practice'. His work was published, for example, in journals of genetics as well as in outlets aimed largely at farmers.[113] And of the two journals which he edited, one of them was a supplement to a weekly magazine for farmers (*Wochenblatt des Landwirtschaftlichen Vereins in Bayern*) and the other, a scientific journal (*Zeitschrift fuer Pflanzenzuechtung*) which he hoped would be able to 'create the necessary link between science and practice'.[114] This Janus-faced orientation is also noticeable in his views on how one should train in order to become a breeder. First of all one should become a competent farmer, he wrote, because without general agricultural knowledge and skill there was little prospect of achieving commercial success in breeding. Then one should

112 Another breeder who belongs in this category is Erich von Tschermak (Harwood, 'Rediscovery of Mendelism').

113 This was Kiessling's claim in a letter to Ministerialdirektor Matt, 25 Jan 1919, BayHStA, MK 35953, and it seems to be true. From 1912 he also edited a series of short volumes written for farmers which dealt with specific technical questions, the *Landwirtschaftliche Hefte* (Schloegl, *Bayerische Agrargeschichte*, 508). Kiessling himself wrote the second volume in the series, a short introduction to the techniques of cereals breeding, which was reckoned to be suitable for use in secondary schools (Fruwirth, 'Ueber den Unterricht'). Moreover his innovative work did not take just written form; of the new pieces of apparatus devised for plant-breeding which the Bavarian station released in the 1920s, Kiessling seems to have invented most of them (Kiessling and Gassner, *Entwicklung und Taetigkeit*, 12).

114 Kiessling, 'Einiges aus der Praxis', 25. (While founder and editor of the *Wochenblatt*, he was on the editorial board of the *Zeitschrift*.) Kiessling's admiration for Kurt von Ruemker was similarly predicated on the latter's ability to link science with practice (Kiessling, 'K. von Ruemker', esp. 2).

acquire the necessary scientific knowledge and become familiar with the general breeding techniques used in the industry. These provided the necessary foundation, but in the final analysis Kiessling felt that successful breeding was more an art than a science:

> Just as instruction in drawing and colour cannot guarantee the development of a great artist, so also are science and technical experience insufficient on their own to produce an innovative breeder [...].[115]

His work at the Plant-Breeding Station also demonstrated his concern to combine the scientific and the practical. Although he had a firm grasp of genetic theory and was regarded by Baur as an outstanding academic breeder,[116] he also championed the interests of Bavaria's peasant farmers. In addition to his breeding work, for example, he undertook a great deal of organisational work, above all setting up a network of farmers' clubs (*Ackerbauvereine*) across the state, on the grounds that the Station's attempts to promote improved varieties and methods would only succeed if the intended clientele was organised.[117]

Much closer to the sceptical pole in the breeding debates was Kurt von Ruemker, professor of agronomy at Breslau from 1895 to 1913. His doubts about Mendelism were part and parcel of his general outlook on breeding which tended to emphasise the importance of practice more than science. On the one hand, he took an early interest in Mendelism, attempting to demonstrate Mendelian laws in barley

115 Kiessling, 'Zur Problemstellung', 20–21.

116 Baur expressed this view to C.B. Hutchison on the latter's visit to Berlin in 1925 (Hutchison papers, box 1, file 33). On the job-offers, see 29 Aug 1919, Kultusminister to Senat der T.H. Munchen, TUM-Archiv, Sign = II, 1, Bd. 5; 24 Dec 1919, Warmbold to Herrn Ministerialrat, BayHStA, MK 35953.

117 The idea caught on quickly, and by 1922 Kiessling and his colleagues at the Station had set up over a thousand such clubs with 32,000 members. Not all existing agricultural organisations welcomed the new arrivals, but Kiessling argued that they were the most appropriate form of organisation for Bavarian conditions (Kiessling, 'Wissenschaftliche und praktische Pflanzenbau-foerderung'). And when conflicts arose between the interests of commercial breeders and those of farmers, Kiessling spoke up on behalf of the latter (Kiessling, 'Einige Gegenwartsfragen', 457).

and wheat and declaring that genetic research would provide the foundation for plant-breeding. But in fact Ruemker seems not to have been particularly well-versed in the new theory.[118] Nor did science loom large in his vision of the breeder: '[...] what a breeder is and achieves cannot be taught; it is the result of a personal gift [...].'[119] His practical orientation was also evident in his views on the organisation of breeding. An admirer of the Swedish plant-breeding station at Svaloef, for example, he praised its 'commercial orientation' and emphasised the fact that it organisation was 'practical through and through'; no one there, he felt, was guilty of 'petty theorising' (*theoretisierenden Kleinlichkeit*).[120] And as far as Germany was concerned, von Ruemker frequently pointed out that since the great majority of farmers were peasants, it was essential that academics and other agricultural experts should devote particular attention to the needs of small farms.[121] Already in 1898 he had called for breeding to be organised on a regional basis and for the peasantry to be involved (a vision which was implemented a few years later with the establishment of plant-breeding stations attached to the Colleges at Hohenheim and Weihenstephan).[122] Just as von Ruemker's views on breeding

118 Von Ruemker, 'Was koennen wir von der Saatzucht', 49; on his Mendelian experiments see von Ruemker, 'Erster Bericht ueber das landwirt. Versuchsfeld'. The British geneticist, William Bateson, met von Ruemker on the former's visit to Berlin in 1914 but was not impressed with his knowledge of genetics (4 Jan 1914, William Bateson to Beatrice Bateson, Bateson papers).

119 Von Ruemker, 'Was koennen wir von der Saatzucht', 49.

120 Von Ruemker, 'Ueber Organisation der Pflanzenzuechtung', 50.

121 Von Ruemker, 'Das landwirt. Versuchs- und Bildungswesen', 111; von Ruemker, 'Ueber Bedeutung und Methoden', 88–89.

122 Von Ruemker, 'Der wirtschaftliche Mehrwert', 148ff. On the Bavarian station see von Ruemker, *Mein Leben*, 246. It may seem puzzling that, despite his admiration for the Bavarian station, von Ruemker – like Baur and Roemer – also believed that public-sector institutions should not produce new varieties for the market as this could only harm the commercial breeding sector (Von Ruemker, 'Erster Bericht', 868–869; von Ruemker, 'Die Entwicklung der landwirt. Pflanzenzucht', 337–338). But there was no contradiction, for although his appeal for regional breeding aimed at the peasantry was taken up by those campaigning for the establishment of *state* breeding stations, the form of organisation which he had actually envisioned was self-help via the formation of cooperatives (von Ruemker, *Die Entwicklung der Landwirtschaft*).

method and organisation were predicated upon a concern with the needs of the majority farming community, so also were his writings on educational reform (cf. Chapter Three) designed to nudge the universities in a more practical direction. And under his leadership from 1895, as we saw above, Breslau became a model institution of this kind.

The other major sceptic who emerges in the breeding debates was Carl Fruwirth. From 1897 until 1907 Fruwirth was professor of agronomy at the College in Hohenheim and from 1905 first director of Wuerttemberg's newly established Plant-Breeding Station.[123] Like Ruemker, he had an initial interest in Mendelism but never came to see it as the breeder's panacea. In a volume of his handbook of plant-breeding which appeared in 1901, for example, he was already discussing Mendelism, and soon after he included a number of Mendelian papers among the works relevant to breeding which he abstracted for a major agricultural journal before the First World War.[124] But Fruwirth's standing among geneticists was not especially high, and when he approached the University of Tuebingen in the hopes of getting his handbook of plant-breeding accepted as a doctoral dissertation, the Faculty of Science objected that it was just a compilation rather than something scientifically novel.[125] Moreover Fruwirth, too, regarded breeding as more of an art than a science:

That his suggestions of 1898 were nonetheless influential upon the movement for state institutions is suggested in Anon., 'Bericht der Koeniglich Bayerischen Saatzuchtanstalt', 1.

123 Boehm, *Biographisches Handbuch*; Wunderlich, 'Carl Fruwirth'; Tschermak, 'Zum 60. Geburtsfeste'.

124 Fruwirth, *Die Zuechtung der landwirt. Kulturpflanzen*, Vol. 1; his abstracts appeared in the *Journal fuer Landwirtschaft*, 1903–1912. He also published an occasional article in journals of genetics.

125 Memoirs of Oskar von Kirchner, UA-Hohen, AZ=N5/1.2 (Kirchner papers), 520. The geneticist, Otto Renner, was not terribly impressed with Fruwirth's discussion of the theoretical foundations of plant-breeding (Renner, [review of Fruwirth]). After meeting Fruwirth during his visit to Europe in the 1920s, the American geneticist, E.W. Lindstrom, ranked him B or C on his 'scale of excellence' in genetics (with A=top and D=bottom) (Lindstrom, 'Notes on Geneticists'). In a letter to the geneticist, William Bateson, Erwin Baur described Fruwirth, similarly, as a 'practical agricultural man' (*praktischer*

> Breeding is still an art even today, for despite an education in the techniques [of breeding] and a familiarity with the laws of genetics, one still requires a breeder's eye in order to be able to spot valuable plants during selection and varietal testing.[126]

As director of Wuerttemberg's station, Fruwirth took his obligations to the region's farmers seriously. If an assistant from the Station was helping a farmer by carrying out a selection, the farmer should have some say in the direction of the selection, he recommended, and it should always be done in the presence of the farmer, so that the farmer could do it alone in future.[127] Realising, however, that few of Wuerttemberg's small farmers would actually wish to take up breeding themselves, he sought to inform the rest of them how they could nonetheless take advantage of the activities of the Station, eg in obtaining the most suitable varieties for their locale and in using simple measures to maintain them.[128]

Fruwirth's successor at Hohenheim, Johann Wacker (professor of agronomy 1907–1933), was also a very practically-minded breeder. Though he occasionally waxed lyrical about Mendelism, Wacker is reckoned by several observers to have been 'closer to practice than to science'.[129] Many of his publications were short articles in the region's

Landwirt) rather than as a scientist (10 Nov 1912, Baur to Bateson, Schiemann Papers).

126 Fruwirth, 'Formen des Pflanzenzuechtungs-betriebes', 647.

127 Fruwirth, 'Formen des Pflanzenzuechtungs-betriebes', 646–647.

128 Fruwirth, *Wie kann sich der Landwirt.* A reviewer noted that this 65-page booklet was clearly written and especially suited for the farmer who just wanted to find out what breeding was and how to use its results (Seelhorst, [review of Fruwirth]). Fruwirth had a lot of experience in writing for a general audience, having been employed as a young man by an agricultural newspaper to answer readers' questions and then teaching for a decade at an agricultural secondary school prior to coming to Hohenheim (Wunderlich, 'Carl Fruwirth').

129 Klein, *Die akademischen Lehrer*; for a similar assessment see Boehm, *Biographisches Handbuch.* At the Station Wacker evidently left the breeding work to an assistant while he concentrated on organising varietal-testing, overseeing varietal certification and other administrative matters (Scheibe, 'Bedeutung der wissenschaftlichen Institute', 65–66). Between 1908 and 1922, correspondingly, I have found no articles by Wacker in the main German genetics journal *(Zeitschrift fuer induktive Abstammungs- und Vererbungs-*

weekly agricultural newspaper (*Wuerttembergisches Wochenblatt*), and coming from a farm background himself, Wacker knew how to reach a farming audience with his writing and public lectures.[130]

It remains to be seen whether the research and development work across the entire range of agricultural disciplines which went on at Berlin, Halle, Bonn, Breslau, and Hohenheim was in accord with the overall orientation of those institutions toward science and practice. What the material in this section does indicate, however, is that the character of the work in *plant-breeding* which went on at those institutions (as well as at Weihenstephan to which we will turn in the next chapter) accorded closely with their general orientations.

Conclusion

As we have seen, 'scientific' and 'practical' institutions differed in their curricula, the extent of their demonstration faciliities, the strength of their ties to various agricultural constituencies, and in the character of their research activity in plant-breeding. The question now is: how are we to account for these differences? Clearly these institutions differed in many respects which might conceivably have affected their orientation toward science and practice, but which ones were in fact influential? Looking at them comparatively has produced a surprise or two. At the beginning of this project, for example, I had thought that those institutions funded by education ministries might be less closely tied to practical concerns than those funded by agriculture ministries. In fact, this seems to have made hardly any difference. Some of the colleges funded by an education ministry were practical in character (Hohenheim, Breslau) while some funded by an agri-

lehre). The one outburst of enthusiasm by him for Mendelism which I have found was his inaugural address as rector at Hohenheim (Wacker, 'Die landwirt. Pflanzenzucht').

130 Wacker, 'Die Entwicklung des Instituts fuer Acker- und Pflanzenbau', 27pp typescript, UA-Hohen, AZ = 3/25.31; Klein, *Die akademischen Lehrer*.

culture ministry were at the scientific end of the spectrum (Berlin). Nor are there any obvious differences in educational policy between the largest states, Prussia and (as we shall see in the next chapter) Bavaria; both maintained some institutions which were 'practical' and others which were 'scientific'.

Similarly, in view of the universities' invocation since the early 19th century of the ideology of 'pure' science, I had originally expected that agricultural education there would have been more scientifically inclined than at the colleges. That hunch was borne out to some extent by the resistance which both Julius Kuehn and Kurt von Ruemker faced in developing the agricultural sciences at Halle and Breslau during the late 19th century, and by the First World War the Prussian Education Ministry seems to have concluded that Philosophical Faculties were not a very satisfactory location for agriculture.[131] But we have now seen that faculty location was not decisive; *some* university institutes did indeed place a heavy emphasis upon the natural sciences while neglecting farm management (Halle), but in other cases where the minister concerned was prepared to override the resistance of the Philosophical Faculty (Breslau), agricultural education was able to develop a practical orientation. The converse was found among the colleges; some were indeed 'isolated' and practical (Hohenheim), but Berlin was neither.

Other features of these institutions' circumstances, however, do appear to have had a certain amount of influence upon their development. One is geographic location. This was particularly evident at Berlin where the College had weak ties to the farming community but was able to take advantage of its proximity to the city's food and drinks industry in order to develop these areas of education and research on a large scale. And while the College's lack of demon-

131 18 July 1919, Kultusministerium to Landwirtschaftsministerium, GStaA, Rep 77, Tit 46, Nr. 31. Friedrich Aereboe expressed this view publically, complaining that candidates whose research was of a basic kind had the advantage in appointments to agricultural chairs in Philosophical Faculties because they were judged largely by scholarly qualifications rather than practical ones (Aereboe, 'Landwirt. Hochschulen oder landwirt. Universitaetsinstitute?'). His claim was inadvertently sustained by Gustav Froelich at Halle who objected that this was *no longer* true (Froelich, 'Nochmals').

stration facilities meant that it could not offer a practically oriented curriculum, the city's cultural amenities, along with the proximity of elite educational institutions like the University, enabled the College to emphasise the 'well-rounded' education and cultivation which students would obtain there. All of this, in turn, also explains why Berlin tended to attract students who had little intention of going into farming.

At the same time, it is important not to exaggerate the explanatory importance of geography. For one thing, while it probably played a role at the extremes of the urban-rural spectrum, it was less significant elsewhere. Institutions situated in small to medium-sized cities, for example – like Bonn (pop. 90,000) or Halle (pop. 200,000) – varied greatly in their orientation to science/practice. Nor is urban-rural setting of much use in explaining why Breslau (pop. 570,000) was a far more practical institution than Munich (pop. 680,000).[132] But even at the extremes of the science/practice spectrum, geography was not as decisive as historical figures often claimed. For although critics accused the urban colleges of turning out 'asphalt farmers' and dismissed the rural ones as being 'hopelessly isolated', the reality was rather different. Since the rural colleges were about a 30-minute train ride away from major cities, both staff and students could enjoy the advantages of living in the metropolis *and* working at well-equipped colleges (as we shall see in the next chapter). Nor were urban colleges *necessarily* devoid of good demonstration facilities, provided the Ministry was willing to pay (eg, Halle or Breslau before the First World War or Berlin in the 1920s). Finally, of course, we should remember that an institution's geographic location was the result, not of an act of nature, but of a decision by the relevant Ministry. Thus geography, though a real material factor in the development of these institutions, could be – and sometimes was – circumvented by human action.

But what kinds of human action were most effective? Let's start with the lobbying activities of academic staff. As we have seen, staff sensitivities about matters of academic status generally prompted them to press for changes in a 'scientific' direction. For example, staff at

132 Population sizes are from *Meyer's Lexikon,* 1925 ed.

Berlin sought to acquire the universities' trappings of prestige; those at Hohenheim wanted new appointees to have doctorates; and at Halle Kuehn favoured scientific rather than economic elements in the curriculum. Whether they succeeded in these campaigns, however, depended entirely on what view the Ministry took. The same point emerges when considering the role played by major entrepreneurial figures. The institute at Halle continued to drift in a 'scientific' direction during the late 19th century because the Ministry made no attempt to put pressure on Julius Kuehn. At Berlin before the First World War, too, those staff complaining about poor demonstration facilities got nowhere as long as the Ministry was in no hurry to move institutes out of the city. Halle's decline was quickly reversed from 1910, however, when the Ministry responded favourably to Ferdinand Wohltmann's requests for a range of new facilities. And at Breslau in the 1890s von Ruemker was able to turn the Institute around only by securing the Ministry's backing for wide-ranging reforms and expansion. Thus energetic leaders with skill and vision were undoubtedly necessary in shifting an institution's orientation, but their success depended upon ministerial willingness to spend money and to knock heads together.

On the face of it, one constituency which might appear to have been influential was students. As we saw in Chapter Three, for example, student organisations were vocal during the educational debates after the First World War. Colleges and universities boasted of large enrollments whenever they could, and institutions faced the threat of closure whenever student numbers declined substantially. Furthermore, insofar as state policy was responsive to the electorate, one can imagine that the geographic origins of an institution's student body might have been significant. Ministers might need to listen more closely where an institution's students came mainly from the region (eg, Breslau) than where they were drawn from much further afield, as at Halle. And a student body recruited from the region would be expected to convey local needs to the government (as well as to academic staff) more forcefully – and thus encourage a more practical orientation – than one which was cosmopolitan in makeup.

Moreover students also varied considerably from one institution to the next in their school and family backgrounds (cf. Chapter Two), and one can imagine that this might have affected an institution's orientation. In the late 1920s, for example, the sons of higher civil servants and big farmers preferred Munich, Berlin, Halle and Goettingen while the sons of lower level civil servants and small farmers were more likely to attend Koenigsberg, Breslau and Hohenheim (the last also attracted a high proportion of the sons of small tradesmen).[133] Thus the student body at scientifically oriented institutions was more likely to be drawn from social elites than it was at practical places. This contrast was particularly noticeable at Halle and Breslau (and we will encounter it again when we turn to Bavaria). Although between 30 and 40% of both Universities' agricultural students were the sons of farmers, two-thirds of those at Halle were from large farms versus about one-quarter of those at Breslau. And while about 15% of the students at both institutes were the sons of higher civil servants, 34% of Breslau's (vs. only 6% of Halle's) were the sons of middle-level civil servants (who would not have had higher education and were generally regarded as belonging to the lower middle-classes).[134] Thus Halle recruited nearly twice as many of its students from these elites as did Breslau. Indeed, Breslau recruited a much higher proportion of its students from non-elite strata (56%) than did any other agricultural institution. This pattern was not confined to agriculture. Contemporaries perceived the University to be more socially open than most, to students from modest backgrounds as well as to Jews.[135]

That this pattern of recruitment should have existed is perhaps not surprising. For it seems likely that elite and 'plebs' sought different things from higher education – a badge of social status for the

133 Bonn lay somewhere in between these extremes. If one looks at the proportion of students coming from elite backgrounds – ie, the sons of big farmers or higher civil servants – Bonn lies about midway between Munich at the top end of the scale and Weihenstephan at the bottom. Nor were Bonn's students as likely to have as extensive schooling as Berlin's (*Deutsche Hochschulstatistik* 1928).

134 Deutsche Hochschulstatistik, 1928.

135 Muensterberg, *The Americans*, 417; Jarausch, 'Universitaet und Hochschule'; Scholz, 'Friedrich Althoffs Einfluss', 342.

former, a technical qualification in the cause of career advancement for the latter. And the class background of the student body would quite likely have served to sustain (or even augment) an institution's orientation toward science/practice. Whether the composition of the student body was itself an independent *cause* of institutional differences of orientation, however, seems unlikely (though I cannot rule it out altogether). For one thing, , the fact that students generally 'vote with their feet' means that most student organisations were unlikely to bother lobbying ministers for very long in order to get changes introduced at any one institution. If it failed to meet their needs, they could usually afford to go elsewhere since the majority came from middle class or large-farm backgrounds.[136] And although it seems plausible that a local student body could serve as a conduit for local concerns, it was evidently of little importance at Hohenheim which maintained a markedly practical orientation despite its cosmopolitan intake.[137] In any event, because of the 'cascade' phenomenon (cf. Chapter Two), it is actually very difficult to find evidence in favour of 'student power'.[138]

By far the most important elements within the fields inhabited by an institution were the ministry which funded it and the major players in the regional agricultural economy. The importance of state policy should be obvious by now since it was state action which granted a college practical facilities (or declined to), adopted (or ignored) staff recommendations when appointing new staff, and upgraded colleges

136 Breslau might be an exception in this respect since its less advantaged students may not have been able to afford to study outside the province.

137 Similarly, despite the differences of orientation between Berlin and Bonn, the geographic pattern of their intakes was quite similar. In 1911 70% of Berlin's students came from: Prussia (vs 80% at Bonn), 16% from elsewhere in Germany (vs 11% at Bonn), and 15% from abroad (vs 9% at Bonn) (Oldenburg, *Das landwirt. Unterrichtswesen in Preussen*). The proportions at both colleges in 1900 were even more similar (*Landwirtschaftliche Jahrbuecher*, 1903).

138 It is interesting that American medical students seem to have been just as powerless. Medical education in the United States drifted steadily closer to a scientific orientation over the 20th century, despite the fact that students consistently favoured a more practical, clinically-oriented curriculum (Rothstein, *American Medical Schools*, passim).

on the academic status-hierarchy.[139] And this helps to explain why the provision for agricultural higher education varied from one college to another within a single state such as Prussia (compare Bonn with Berlin or Breslau with Halle) or Bavaria (as we shall see in the next chapter).

Another major force within the politico-economic field, of course, was the farming community. In the case of Hohenheim, we saw how Wuerttemberg's Chamber of Agriculture was able to block staff attempts to introduce a rectorial constitution and to abolish the advisory board. And at Breslau both the Chamber and the Agricultural Association provided crucial support, both financial and political, which enabled von Ruemker to develop the Institute in a practical direction. But as we have seen, the particular *composition* of the farming community was crucial. Among the exemplary institutions, for example, those of a practical orientation were to be found, by and large, in regions dominated by small farms: Bonn, Hohenheim (and Weihenstephan as we shall see in Chapter Five).[140] 'Scientific' institutions like Berlin and Halle were located in regions which had a relatively large number of big farms. This meant that academics at those institutions had access to a very different constituency and one which could be expected to take an interest in research and curricula of a more 'scientific' kind.

Finally, it is clear from the exemplary institutions that farmers were not the only influential interest group. This was most noticeable at Berlin where various branches of the food and drinks industry had a substantial impact on the College while Erwin Baur addressed his plant-breeding work to the Berlin-based association of commercial breeders *inter alia*. In Halle, similarly, Theodor Roemer's institute

139 Again, medical education has been similar: 'Governments have been responsible for every significant innovation in medical education since [the] mid [20th] century' (Rothstein, *American Medical Schools* , 334).

140 To be sure, Breslau is an exception since Silesia had a relatively large number of farms over 100 ha (cf. Table 2.1). But this did not prevent von Ruemker from taking a particular interest in making the results of plant-breeding available to small farmers. And as we have seen, the presence in the city of major farming organisations, combined with the University's tradition of service to the region, tended to channel agricultural education in a practical direction.

cultivated close ties to the region's sugar and fertiliser industries as well as to its many commercial breeders.

While many factors played a modest role, therefore, the development of educational institutions was primarily shaped by the interaction of a small number of countervailing forces. Staff in search of academic respectability tended to shift institutions in a 'scientific' direction, a trend which could be facilitated or hindered by the two major players in the politico-economic field: state agencies and agricultural interest groups. Although this model goes some way toward accounting for the orientations of these five exemplary institutions, however, it still leaves out an important dimension. For in attempting to provide a synchronic snapshot of each institution during the decade or two before the First World War, I have largely bracketted out its *history*, even though the decisions taken about an institution during that period, whether by ministers or by academic staff, are often only explicable in terms of prior events or developments. Remedying this deficiency is not easy since the existing secondary literatures on the history of these institutions are, for the most part, celebratory and thus do not address the key analytical issues. This has meant that the only way to recapture the relevant past is to turn to the primary materials and 'do it yourself'. That is the task of the next chapter where we will look in detail at the longer term evolution of two contrasting institutions within a single state, each of them competing for academic as well as political legitimacy.

Chapter 5: Manoeuvring for Position: Bavaria's Agricultural Institutions, 1872–1934

Bavaria provides an ideal setting in which to follow the evolution of higher agricultural education over a longer period. For one thing, since it possessed only two such institutions – at Munich and Weihenstephan – which displayed sharply contrasting orientations toward science/practice into the 1920s, comparative analysis is relatively easy. Moreover, the available source material is rich and diverse. In addition to the usual primary materials in state and university archives, the secondary literature on the history of Bavarian agricultural education is relatively extensive, much of it written by an insider who witnessed first-hand some of the events of the interwar period (Hans Raum, professor of plant production at Weihenstephan). Finally since these two institutions were accountable to the same ministry, their histories were closely intertwined. And this fact allows me to convey more concretely what it means to say that an institution develops within an evolving 'field' of possibilities defined in part by the position of other competing academic institutions as well as by the presence of potential clienteles which can be mobilised for support.

The origins and early history of Munich and Weihenstephan

Weihenstephan's origins date from 1803 when it was established as an agricultural secondary school on the site of a monastery outside Freising, a cathedral town north of Munich. Like its engineering counterparts, the 'polytechnic schools', Weihenstephan was originally overseen, not by the Education Ministry but by that for Trade. In effect, it was a boarding school for the sons of German and foreign estate-owners, and its staff were expected by the Ministry, not only to educate its pupils (called 'eleves' though they were generally aged between 18 and 22), but to maintain strict discipline. Like the academies elsewhere in Germany, Weihenstephan was attached to a state farm and expected to provide a practical form of instruction. Sensitive to Liebig's critique of the academies during the 1860s, Weihenstephan's director attempted to strengthen its sciences by asking for the establishment of a post in agricultural chemistry there, along with the transfer of an agricultural experiment station from Munich to Weihenstephan. Significantly, the Ministry rejected his application on the ground that Weihenstephan was a 'practical and teaching' institution, whereas science (*Wissenschaft*) needed to be situated near the University and the Academy of Sciences in Munich.[1]

Since the late 18th century agriculture had been taught at Bavarian universities as part of the administrative sciences (viz. at Wuerzburg and Munich). At Munich the subject continued to be taught through most of the 19th century (usually by professors originally trained in chemistry or botany).[2] On the last incumbent's death in 1875, however, the chair was not refilled because agriculture had recently been established at the city's engineering college. In 1872 the Bavarian parliament had voted to establish a 'Faculty (*Abteilung*) of Agriculture' at the college which would provide 'the highest theoretical scientific instruction'.[3] The proposal for such a Faculty had apparently come from the Agricultural Association of Bavaria (*Landwirtschaftlicher Verein in Bayern*) whose agricultural experiment station was incorporated into the new Faculty at the outset. The Faculty also enjoyed the support of Justus von Liebig who had been professor of chemistry at the University since 1850 and whose influential 1861 speech denouncing the academies had been delivered at the Bavarian Academy of Sciences. Although doubts were raised at the time whether Bavaria needed a second agricultural institution, the Bavarian government replied that Munich and Weihenstephan would be offer-

1 Steuert, *Die Koenigliche Bayerische Akademie*; the phrase is at 173–174.
2 Raum, 'Die akademischen Lehrgestalten'.
3 Jenne, *Das landwirtschaftliche Unterrichtswesen*, 25.

ing very different kinds of education.[4] By deciding to establish a second institution at the engineering college, however, the government was not merely introducing a new and complementary form of education; in effect it was also lowering Weihenstephan's academic status.

If the aim had been to attract a new group of students to agriculture, it seems not to have worked. Student recruitment at Munich was poor throughout its first decade, though numbers also declined at Weihenstephan.[5] In the early 1880s, therefore, a committee of the Bavarian legislature proposed closing the Faculty in Munich because of its failure to recruit (a threat which was to recur in future). Ministers resisted but agreed to consider reducing the apparent competition between the two institutions by establishing a clearer division of labour between them, in which the curriculum at Munich would henceforth be made more 'scientific' and that at Weihenstephan, more 'practical'. Staff in Munich were not convinced that such a division of labour would solve their problems. They felt isolated among the much larger engineering and science departments at the engineering college and were very aware that their institutional location was rather odd within German higher agricultural education. The only long-term solution, they argued, would be for agriculture to merge with the forestry institute in the Faculty of Public Administration (*Staatswirtschaftliche Fakultaet*) at the University. This would have been an institutional solution similar to that which von Ruemker was to propose at Breslau a few years later, and it was a move which would have reduced Munich staff's isolation while enhancing their academic status. But the proposal got nowhere since the engineering college was not prepared to bid farewell to agriculture.[6]

4 Raum, [untitled collection].

5 Into the early 1880s Munich had on average only 8 regular students plus half a dozen part-timers while Weihenstephan had about 40 regulars, though the latter number fell over the next decade (Raum, [untitled collection], 12).

6 Anon., 'Kurze Darstellung der Entwicklung der landwirt. Zentralschule [...].', 18 page typescript enclosed as appendix 16 with letter of 6 May 1926, Kultusminister to members of Sachverstaendigenkommission, BayHStA, MK 19682. Raum, 'Die landwirt. Abteilung der T.H. Muenchen'.

Low student numbers continued to plague both institutions during the 1880s, and the issue of reorganisation did not go away. In 1891 (and in 1894) Munich's agricultural staff once again proposed that they should move to the University and that their institutional future should be secured by closing Weihenstephan.[7] But the proposals to close the latter failed, partly because of the appointment in 1892 of a new director there, Carl Kraus, who soon reversed its deteriorating position. With the full backing of staff, Kraus introduced a variety of improvements with the aim of upgrading Weihenstephan from a secondary school to an academy. By terminating the provision of on-campus accommodation for pupils, space was freed for new institutes. Steps were also undertaken to develop the state farm as a research facility rather than simply as a profit-making enterprise used for teaching demonstrations.[8] Suspecting that staff in Munich would be hostile to Weihenstephan's upgrading, Kraus insisted that the two institutions would remain very distinctive in character and would thus complement one another. It might even make sense, he suggested, for them to enter into association as a single college with two diverse branches.[9] (In terms of the model outlined in Chapter Two, Kraus was pushing for the upgrading of Weihenstephan – by developing its research capacity – though without moving it in the more 'scientific' direction so often associated with academisation since his options were limited. To have gone down the 'scientific' route would have constituted an outright challenge to the higher status institution in Munich and would have flown in the face of the Ministry's principle of a division of labour.)

In stressing the institutions' complementarity, Kraus presumably hoped to avoid provoking opposition to his manoeuver from staff in Munich, but in the event he underestimated their sensitivities. Why turn a perfectly good secondary school, Munich's staff countered, into an academy when Bavaria already had sufficient provision for higher agricultural education? Besides, Weihenstephan's curriculum was so

7 Soxhlet, 'Denkschrift'.

8 Steuert, *Die Koenigliche Bayerische Akademie*.

9 Raum, 'Weihenstephan unter Lintner und Kraus'; for Kraus's complementarity argument see Kraus, 'Die Pflege'.

narrowly utilitarian: 'How can the intellectual life of our people advance if new generations are educated merely for earning a living?'[10] With these arguments, however, Munich was conveniently turning a blind eye to the widely acknowledged fact that Weihenstephan was by then far better equipped with demonstration facilities. Indeed, in the years before the First World War staff at Munich often made arrangements to take their students to Weihenstephan for practical demonstrations.[11] Not satisfied with trying to block Weihenstephan's proposal for upgrading, Munich also tried to widen the status differential between the two institutions by bidding upwards: in 1896 they lobbied once again for a merger with the University's forestry institute.[12]

In the event, neither of the mergers went ahead. But in 1895 the Ministry accepted Kraus's bid, agreeing to upgrade Weihenstephan to an academy and to improve its facilities substantially. The teaching staff was expanded with two new professors and two instructors,[13] and the curriculum was modified so as to contain more basic sciences (eg, plant nutrition and physiology). Recurrent state expenditure on Weihenstephan, having been somewhat less than Munich's during the 1890s, now leapt ahead, approaching twice Munich's level by 1905.[14] In addition, the establishment of a large number of new agricultural secondary schools (*Winterschulen*) during the 1890s meant that there was a growing demand for teachers, and Weihenstephan was quickly inundated with students.[15]

By now perceiving Weihenstephan as a serious competitor, Munich complained bitterly of its historical underfunding, but there was no response from the Ministry, and in the Bavarian parliament the question of Weihenstephan's relationship to Munich continued to be

10 Jenne, *Das landwirtschaftliche Unterrichtswesen*, 46.

11 *Bericht ueber die T.H. Muenchen*, 1901/02–1913/14; Jenne, *Das landwirtschaftliche Unterrichtswesen*.

12 Wollny and Pott, 'Antrag betreffend die Verlegung'.

13 Raum, 'Weihenstephan unter Lintner und Kraus'.

14 Jenne, *Das landwirtschaftliche Unterrichtswesen*, 42–44, 285.

15 Raum, 'Die akademischen Lehrgestalten'.

raised. [16] In 1908 the Faculty's fortunes seemed to have brightened when it acquired more spacious accommodation in a wing of the new main building of the engineering college (though still without demonstration facilities).[17] But in 1909 staff at Weihenstephan proposed a further partial upgrading which would give them some but not all of the features of a college, and their appeal found support in the parliament. Seeing this as the thin end of the wedge, Munich reacted sharply, saying that this proposal overlooked the fact that Weihenstephan was a 'distinctive kind of institution with its own aims'; upgrading would be a regressive move since it would place both institutions in direct competition with one another.[18] Fortunately for Munich, the Education Minister defended the status quo, saying that Bavaria had no need of two institutions of college rank. When some parliamentarians countered that if two were too costly, there must be a case for closing down Munich, staff at Munich insisted that their 'scientific' (*wissenschaftlich*) research and education were an essential complement to Weihenstephan's different goals and stronger 'practical' emphasis. The issue refused to go away, and when yet another petition for Weihenstephan's upgrading was tabled in 1910, the Minister agreed to convene a commission to look into the relationship between the two institutions.[19] In the event, proposals to close Munich down were defeated, but on the eve of the First World War it was common knowledge in agricultural circles that Munich's future was uncertain.[20]

16 On underfunding see Soxhlet, 'Denkschrift'. On parliamentary questions see *Muenchener Neueste Nachrichten*, 8 Mar 1898 (transcript in TUM-Archiv, Sign = I, 8, Bd 1); 12 July 1906, stenographed report of the 172nd sitting of Bavarian Kammer der Abgeordneten, pp 359–360; and 23 June 1908, report of 150th sitting, pp 581–590 (both in BayHStA, MK 19678).

17 *Bericht ueber die T.H. Muenchen*, 1908–09.

18 17 Nov 1909, Vorstand der Landwirt. Abteilung der T.H. Muenchen [Kraus] to Senat der T.H. Muenchen, BayHStA, MK 19679.

19 On Weihenstephan's 1909 appeal to upgrade their status see Direktion der Akademie Weihenstephan to Staatsministerium des Innern, 26 June 1909; see also 19 July 1910, report of 354th sitting of Kammer der Abgeordneten, pp 481–487 (both in BayHStA, MK 19679).

20 Ruemker, 'Organisation und Ziele', 29; Anon., 'Kurze Darstellung der Entwicklung der landwirt. Zentralschule [...].', 18 page typescript enclosed as

Munich and Weihenstephan compared, ca. 1920

To the casual visitor, the college at Weihenstephan and the Faculty in Munich would certainly have appeared very different. Situated 40 km north of Munich in open farmland on the outskirts of the town of Freising, Weihenstephan was an independent college of agriculture (with a famous school of brewing which occupied the site of a medieval monastery there). The Faculty, on the other hand, constituted the smallest division of the engineering college in Bavaria's largest city. But it was not just their physical settings which were radically different.

The first thing which prospective students might have noticed was that the two places offered different kinds of degrees. The only degree which one could take at Weihenstephan (until 1922) was the four-semester 'practical' course aimed at those who would go on to own or manage large estates (*Lehrgang fuer den praktischen Landwirt* or '*kleiner Studiengang*'). At Munich, on the other hand, from 1880 the only agricultural degree on offer was the six-semester programme designed for those who intended to go into teaching in agricultural secondary schools or into administrative jobs in government or private-sector agricultural organisations. In fact, by the late 1920s Munich was the only agricultural institution in Germany which still did not offer the four-semester degree (to have done so would presumably have undermined its claim to offer an education distinct from Weihenstephan's).[21] As we saw in Chapter Two, the entry-qualifications required for these degrees were different: more academic for the six-semester programme, more practical for the four-semester one. As a result by the late 1920s 85% of the students at Munich had the *Abitur* while only a third of those in Weihenstephan did (a fact which undermined Weihenstephan's academic status). In addition, Munich was then the only institution in the country which did not require prior

appendix 16 with letter of 6 May 1926, Kultusminister to members of Sachverstaendigenkommission, BayHStA, MK 19682.

21 Raum, 'Die landwirtschaftliche Hochschullehre'; Jenne, *Das landwirtschaftliche Unterrichtswesen*; Schroeder, *Das Studium der Landwirtschaft*.

practical experience for entry.[22] In terms of their degree-programmes, therefore, one could not have found two more distinct agricultural institutions in Germany.

The curricula at Weihenstephan and Munich were also quite different. At Munich about 1905, for example, students took basic sciences and economics in their first year and did not encounter any specifically agricultural courses until the second year. This sequence made sense, according to a petition calling for expansion at Munich, because 'in its technical component agriculture is an applied science.'[23] At Weihenstephan, however, students in their first semester took courses in basic sciences and economics but also in a range of agricultural subjects (eg, plant pathology, agricultural machinery and animal husbandry).[24] In consequence, students at Munich spent 50% more time on physics, twice as much time on mineralogy and geology, and more time on building science – but less on laboratory exercises in plant pathology or on horticulture – than did those at Weihenstephan.[25] When the science-minded American agricultural scientist, C.B. Hutchison, visited both institutions in 1925, he came away with a favourable impression of Weihenstephan but felt that 'perhaps [its] greatest weakness [...] is a tendency to overemphasise practical instruction.'[26]

22 On students with the *Abitur*: *Deutsche Hochschulstatistik*, 1928. On practical entry-requirements: Schroeder, *Das Studium der Landwirtschaft*.

23 'Aufruf an die aussenbezeichnete Adresse', n. d. but probably Dec. 1922, TUM-Archiv, Sign = I,8,Bd. 1. The agricultural sciences, as one professor put it, were taught as 'applications [...] of the basic sciences' (Henseler, 'Die landwirt. Abteilung der T.H. Muenchen', 411).

24 Jenne, *Das landwirtschaftliche Unterrichtswesen*. The same was true in 1910–11, though dairy science displaced plant pathology *(Bericht-Weihenstephan*, 1910–11). This pattern persisted into the 1930s: see 'Studienplaene: Pflichtfaecher des Diplom-Studienganges' and 'Pflichtfaecher des Studienganges fuer praktische landwirte', TUM-Archiv, Sign = II, 1, Bd. 4.

25 'Zur Gegenueberstellung der Pflichtfaecher in Muenchen und Weihenstephan laut Programm 1928–29', ca Oct/Nov 1929, BayHStA, MK 19683.

26 p.190, file 33, box 1, C.B. Hutchison papers. Jenne had earlier also perceived Weihenstephan's curriculum to be 'excessively' practical (Jenne, *Das landwirtschaftliche Unterrichtswesen*, 45–46).

As we have seen elsewhere, the basic sciences were also taught differently at these two institutions. In Weihenstephan basic sciences were taught by the college's own staff and were tailored to the concerns of agriculture students.[27] In Munich the courses in basic sciences taken by agriculture students were the same ones provided for engineering students and were taught by the science staff of the engineering college. The director of the Agricultural Faculty defended this arrangement, arguing that basic sciences should be taught for their own sake, rather than on behalf of agriculture, because one could not know in advance which bits of basic science might eventually prove to be applicable.[28]

Another thing which would have struck visitors to Weihenstephan was the variety of its demonstration facilities. Like the other former academies, Weihenstephan had a model farm (280 hectars), run on a commercial basis but used for teaching and research, as well as experimental fields (75 and 8 hectars) for agricultural chemistry, agronomy and cultivation technology. Several state experiment stations were also located in Weihenstephan and attached to the College: the plant-breeding station, an institute of dairy science (which included an experimental dairy, experimental laboratories, and experimental dairy farm), a testing-station for agricultural machinery, and a station for moor-cultivation. It also had a state dairy school and a horticultural school.[29]

The contrast with Munich was stark. To be sure, the Faculty benefited from the proximity of one or two state research and testing-stations which were situated in Munich. For example, the director of the Royal Agricultural-Botanical Station (*Kgl. Agrikultur-Botanische Anstalt*, est. 1902) taught agricultural bacteriology part-time at the

27 Hutchison found the basic sciences and economics as taught in Weihenstephan to have a 'distinctly agricultural slant' (p 260, file 33, box 1, C.B. Hutchison papers).

28 See Kiessling's remarks in 13 Oct 1925, minutes of meeting of sub-committee, BayHStA, MK 19682; cf. Kiessling, 'Die Stellung der Landwirtschaftslehre', 8.

29 Kulisch, 'Gesichtspunkte zur Entscheidung ueber den Ausbau des landwirt. Hochschul-Unterrichtes in Bayern', 17 Jan 1923, BayHStA, MK 19682; *Bericht-Weihenstephan*, 1912–13; Hilz, 'Weihenstephan: Landwirte'.

Faculty.[30] Though an agricultural experiment station had also been part of the Faculty from its founding, until 1905 it possessed no experimental field. The Faculty's institute of agronomy did possess an experimental field, but it was only 1 hectar, and half of that was uncultivatable, as the institute's director complained.[31] When C.B. Hutchison visited in the 1920s he remarked that the Institute's experimental plot was so small that most research had to be conducted at the plant-breeding station in Weihenstephan. Similarly, the institute for animal husbandry had neither a farm nor animals, so experimental work relied on cooperation from local farmers.[32] To be fair, the fact that Munich lacked a model farm did not distinguish it from most university-based institutes, nor was it any worse off than Berlin. But those students who transferred to Munich after studying at Weihenstephan were not impressed, and they soon found themselves being mockingly referred to as 'asphalt farmers'.[33] The dearth of facilities meant that instruction at Munich had to rely on lectures and laboratories except when students could be taken to Weihenstephan for demonstrations and exercises.[34] Trying to play down this deficiency, Munich's staff sometimes argued that hands-on practical knowledge was better obtained through work experience before entering higher education.[35]

Unfortunately for Munich, it was not just its demonstration facilities that fell short. While Munich had four full-time staff in agri-

30 *Bericht ueber die T.H. Muenchen*, 1902–03.

31 15 July 1927, Kiessling to Rektor of T.H. Muenchen, TUM-Archiv, Sign=V, 33, Bd. 2.

32 log section xxvii, 'Germany (3rd report) 1925', box 2, C.B. Hutchison papers.

33 Raum, 'Lebenserinnerungen', 9–11; similarly Haushofer, 'Bauer im Schach', 20. On asphalt see: Dr. Schmauser, Denkschrift, 29 Dec 1929, BayHStA, MK 19684; and Senat der Technischen Hochschule to Kultusminister, 26 Mar 1926, BayHStA, MK 19682.

34 Kulisch, 'Gesichtspunkte zur Entscheidung ueber den Ausbau des landwirt. Hochschul-Unterrichtes in Bayern', 17 Jan 1923, BayHStA, MK 19682; Jenne, *Das landwirtschaftliche Unterrichtswesen.*

35 eg, Kiessling's remarks in minutes of meeting at Kultusministerium on 22 Nov 1929, BayHStA, MK 19683; Theodor Henkel to Senat der T.H. Muenchen, 28 Nov 1922, TUM-Archiv, Sign = I, 8, 1.

cultural subjects by 1910, Weihenstephan had seven.[36] Of course, size isn't everything. It is important to know how these staff were educated and what other kinds of qualification they possessed.

Table 5.1 – Educational and Career Backgrounds of Staff at Munich and Weihenstephan, 1900–1920[37]

	Munich	*Weihenstephan*
attendance at classical grammar school[38]	5	7
doctorate at appointment	7	7
Habilitation	2	0
practical experience pre-appointment[39]	4	9
native of Bavaria	1	10
Staff totals	7	10

36 Raum, 'Die Ereignisse', 17; *Bericht-Weihenstephan*, 1910–11; data on T.H. Munich were pieced together from a variety of sources.

37 The data have been compiled from Raum, 'Die akademischen Lehrgestalten'; Raum, 'Die landwirt. Abteilung der THM'; Bergmann, *Bayerischer Bauernbund*; and from personal files at Archiv-Weih, TUM-Archiv and the Bavarian State Archive. The staff at Munich are Carl Leisewitz, Martin Wollny, Franz Soxhlet, Emil Pott, August Schnider, Hans Henseler and Leonhard Vogel. The staff at Weihenstephan are Carl Kraus, Heinrich Puchner, Joseph Ahr, Ludwig Kiessling, Theodor Henkel, Josef Spann, Carl Kronacher, Anton Fehr, Hans Rauch and Friedrich Wagner. Several staff transferred during this period from one institution to the other (Kraus, Henkel and Kiessling, all from Weihenstephan to Munich). Rather than counting them twice, I have treated them as belonging to their original institution, Weihenstephan.

38 ie, humanistisches Gymnasium.

39 ie, work experience in agriculture between completing higher education and taking up an academic appointment.

Table 5.1 compares the full-time staff (in agricultural disciplines only) who worked at these institutions at some point during the period 1900 through 1920. We see that in terms of schooling Munich and Weihenstephan were quite comparable; most of the staff at both institutions had attended a classical grammar school. In terms of higher education, however, staff at Weihenstephan were noticeably less qualified. Though all of the staff at Munich had doctorates by the time of their appointments, three of those at Weihenstephan did not (Joseph Ahr, professor of agricultural chemistry, Hans Rauch, professor of agricultural accounting and commercial studies, and Anton Fehr, professor of dairy science). Weihenstephan's staff were not happy about appointing Fehr without a doctorate, perhaps fearing that it would be used to ridicule the College (and it was).[40] And although a few of Munich's staff were habilitated, none of Weihenstephan's were. The contrast between staff at the two institutions is sharper, however, if one considers their *non*-academic qualifications. Between completing their higher education and taking up their first academic post, 90% of the staff at Weihenstephan had held non-academic jobs – as teachers in agricultural schools, administrators in agricultural organisations, farmers or estate-managers, veterinarians or as technical staff in agricultural industry – versus slightly more than half of those at Munich. Thus the educational and career trajectories of staff over this period sustain the picture emerging above, that Weihenstephan was a far more practically-oriented institution than Munich.

40 Raum, 'Die von 1951–1972', 3; Direktor of Weihenstephan (Vogel) to Kultusminister, 6 Nov 1919, 1 Dec 1919 and 19 May 1920; Lehrerrat Weihenstephan to Kultusminister, 28 Nov 1919. Hans Rauch was appointed as professor and head of agricultural accountancy (*Buchstelle*) in 1917 on the understanding that he would get his PhD (Vogel to Kultusminister, 19 May 1920). The head of the Brewing Faculty, Theodor Ganzenmueller, also had no doctorate (Vogel to Kultusminister, 19 May 1920) (all correspondence from BayHStA, MK 19697); and *Bericht-Weihenstephan* for 1913–14, 94. Thus Raum's claim that staff at the two institutions were identical in their education and academic qualifications (Raum, 'Die landwirt. Hochschullehre', 481) was mistaken. For the Director's concern that Weihenstephan's academic reputation would be harmed by appointing staff lacking a broad scientific education or doctorate, see 16 Nov 1907, Direktor of Weihenstephan to Kultusminister, BayHStA, MK 19703. In the event, he did not get his way.

That staff backgrounds should have differed in this way is not surprising if one considers the appointments process. For it is clear that appointments committees at the two institutions were applying somewhat different criteria. Consider, for example, the appointment of a professor of animal breeding at Weihenstephan in 1917. The committee sought a candidate who combined 'scientific training and practical experience'.[41] One of the candidates, Wilhelm Zorn, had the support of a local cattle control association (*Rindviehkontrollverein Steinach*) who praised him for his 'practical' familiarity with the latest developments in breeding. 'Only [a candidate] who has practical and not *just theoretical experience* in *all branches of agriculture* and who has long years of experience as a breeder', the Association declared, could adequately fill this post.[42] However another referee objected that Zorn was completely ignorant of the state of animal breeding in Bavaria (just as ignorant, the referee added, as Zorn's mentor, the ex-professor of animal breeding at Munich, Emil Pott). Moreover Zorn's views of breeding practice would make him many enemies among Bavarian farmers. This referee's preferred candidate was Josef Spann who was said to be intimately familiar with Bavarian conditions, had published a good deal, and had 'mastery of the whole field of animal breeding, both practically and scientifically'.[43] Judging from Spann's list of publications, he appears to have been very much a practically-oriented scholar. An editor of the magazine, *Alm und Weide* (alm and pasture), his publications were all concerned with either practical breeding matters or alpine agriculture rather than with more 'scientific' issues (such as the relative efficacy of various breeding practices).[44] Believing it to be crucial that the appointee should be able to maintain good relations with Bavarian breeders, the committee opted for Spann.

41 25 Aug 1917, Kultusminister to King, BayHStaA, MK 19696.

42 20 June 1916, transcript of deposition from Rindviehkontrollverein Steinach to Ministerialrat Bader, BayHStaA, MK 19696 (emphasis in original).

43 25 Feb 1917, 'Referat 14', signed by Attinger, BayHStA, MK 19696.

44 'Abhandlungen ueber Tierzucht, Alp- und Weidewirtschaft veroeffentlicht von Tierzuchtinspektor Dr. Spann', BayHStA, MK 43314, Personalakte Josef Spann.

By contrast, consider an appointment in the same field at Munich a few years later. Here, too, among the committee's declared criteria were 'scholarly significance' (*wissenschaftliche Bedeutung*) and 'practical experience'.[45] When we look at the arguments deployed for and against particular candidates, however, it is evident that these two criteria were not regarded as of equal value. To be sure, where candidates had extensive practical experience of agriculture or knowledge of Bavarian conditions, this was cited as a plus, and one member of the committee placed particular emphasis upon these qualifications.[46] But overall the committee's sympathies lay elsewhere. For example, one candidate (W. Voeltz) received support from a member of the committee (as well as from the departing professor of animal breeding) despite the fact that Voeltz evidently lacked knowledge of either practical animal-breeding or Bavarian conditions.[47] Another candidate (Dettweiler) was criticised on the ground that his work was 'too practical'. The committee's first choice (who was subsequently appointed) was Heinz Henseler, a man said to have had no knowledge of Bavarian conditions. Of particular interest is the way in which Henseler was portrayed in the committee's report to Senate in which they set out the arguments for their short-list.[48] Although Henseler was in fact the son of a farmer and had practical experience of agriculture, neither fact was mentioned in the report. Instead his strengths were said to be his reputation, excellent training in animal-breeding, talents as a teacher, and the high quality of his publications (the last of these was underlined). In Munich, it appears, scientific qualifications counted more heavily.

If these two appointments were typical, it would be easier to understand the most striking contrast in Table 5.1: that whereas all of

45 17 Jan 1920, minutes of meeting of Lehrer-Kollegium, TUM-Archiv, Sign= II, 1, Bd. 5.

46 It may be significant that the member in question was Theodor Henkel, an agricultural chemist who had moved to the Munich Faculty a few years earlier after a decade at Weihenstephan.

47 17 Jan 1920, minutes of meeting of Lehrer-Kollegium, TUM-Archiv, Sign= II, 1, Bd. 5.

48 2 Feb 1920, August Schnider to Senat der Technischen Hochschule, TUM-Archiv, Sign = II, 1, Bd. 5.

the staff at Weihenstephan were natives of Bavaria, only one at Munich was. And Bavarians were presumably more likely than outsiders to have the knowledge of local agricultural conditions deemed so important at Weihenstephan. But were these appointments typical? Consider the appointment of a professor of agronomy at Munich in 1919. In presenting its short-list to the Senate, the appointments committee placed heavy emphasis upon scientific excellence. Their first choice (E.A. Mitscherlich), they pointed out, had received a prize from the Bavarian Academy of Sciences for his *Habilitation* dissertation and had enjoyed an extraordinary degree of success as a researcher. The number two candidate (Heinrich Puchner), the committee felt, had published a substantial number of 'valuable scientific works', but although Mitscherlich's equal in teaching ability and teaching experience, he could not compete in terms of the quality or quantity of his scientific achievements. The third candidate (Joseph Ahr) was highly respected as an advisor to farmers, and his publications in plant husbandry were of great practical value, but the committee reckoned his work had not produced any fundamentally new ideas.[49] In its reply to the Faculty's nominations, however, the Ministry of Education – probably under pressure from the farming community – asked the Faculty to reconsider its short-list. A number of spokesmen from agricultural circles had expressed the desire that the appointee should have extensive knowledge of, and close ties to, Bavarian agriculture, and they believed that Ludwig Kiessling (then a professor at Weihenstephan) was the best man for the job.[50] The appointments committee immediately responded, acknowledging Kiessling's organisational talents, his excellent work as director of the Bavarian Plant-Breeding Station, and his contributions to practical plant-husbandry. But as a plant-breeder, the committee insisted, Kiessling was too specialised (*einseitig*) for a chair of agronomy. Thus they stuck to their original short-list, adding: 'One should not underestimate the advantage of academics maintaining close contact with

49 31 Jan 1919, Landwirtschaftliche Abteilung to Senat der Technischen Hochschule, TUM-Archiv, Sign = II, 1, Bd. 5.

50 29 Aug 1919, Kultusminister to Senat der Technischen Hochschule, TUM-Archiv, Sign = II, 1, Bd. 5.

farmers, but this factor ought not to be decisive.'[51] In saying this Munich's staff was placing scientific credentials before practical ones (as well as distancing itself from practical institutions: all three of the candidates whose scientific accomplishments they found less impressive – Puchner, Ahr and Kiessling – were at that time employed at Weihenstephan). Nevertheless, the Minister's wishes prevailed, and Kiessling was appointed.[52] Once again we see that academic staff were successful in realising their aims only so long as the state (or the interest-groups to which it was responding) condoned them.

For comparative purposes it would have been nice to look at an appointment in agronomy at Weihenstephan, but the necessary archival materials are not available. Let us, therefore, consider instead the way in which staff at Weihenstephan went about filling the chair of agricultural chemistry in 1910.[53] At the outset the committee declared it more important to appoint a practically-inclined man (*Praktiker*) who had contacts with influential figures in agricultural circles and would thus be able to restore Weihenstephan's reputation (which they reckoned had declined in recent years) than to focus upon candidates' competence as chemists. Unfortunately, none of the applicants for the post met that requirement. The committee was unanimous, however, that the best man for the job was Josef Ahr – though, as we saw above, he had no doctorate – and Ahr was duly appointed.

Appointments committees at the two institutions thus placed rather different emphases upon ties to the farming community. But what kinds of research and extension work did staff actually carry out once appointed? When Weihenstephan was upgraded to an academy

51 29 Aug 1919, Landwirtschaftliche Abteilung to Senat der Technischen Hochschule, Bay HStA, MK 35953.

52 22 Jan 1920, Kiessling to Ministerialdirektor Matt, BayHStA, MK 35953, in which Kiessling expresses his thanks for Matt's support in the appointment. It would be wrong to suggest that Munich's appointments committees *invariably* weighted scholarly qualifications higher than practical ones. This was not so, for example, in their appointment of a professor of farm management in 1904 (MK 19564, BayHStA), but among those appointments proceedings I have seen, this latter case was atypical.

53 5 Jan 1910, Direktion der Akademie [...] Weihenstephan to Staatsministerium des Innern, BayHStA, MK19696.

in 1895, the opening paragraph of its new constitution spelled out the institution's overall responsibilities:

> Through its scientific and practical equipment, as well as through the corresponding activities of its staff, the [Academy] is to contribute wherever possible to the theoretical and practical promotion of agriculture and brewing. It is to provide an advisory service for Bavarian farmers and brewers, to encourage them to undertake experiments, and to carry out tests on new machinery.[54]

This appears not to have been empty rhetoric. Weihenstephan's annual reports between 1899 and 1914 are peppered with details of extension work: lectures given to agricultural associations, expert testimony provided to state and private agricultural organisations; visits from agricultural schools, clubs, and farmers' groups; staff serving as judges at agricultural shows, and the provision of a variety of short courses (in 1912–13, for example, there were courses on bee-keeping, hops-testing, horticulture, dairying and animal husbandry). Staff also provided written replies to farmers' questions and carried out tests on samples which had been sent in. Similarly, the Accountancy Centre (*Buchstelle*) was established in 1909 to provide farmers with book-keeping advice or calculations of profitability and to conduct courses on these matters, and in 1923 the institute of farm management also established an advisory office for farmers.[55]

If one reads the annual reports from the Faculty at Munich, on the other hand, contact with Bavarian farmers was evidently more limited.[56] Lectures to various agricultural organisations (including Bavarian ones) were quite common from 1901, as was expert testimony (*Gutachten*) by the professors of farm management, and occasionally staff served as judges in agricultural shows. By far the most common form of contact consisted of staff taking students on excursions (by no means always in Bavaria) to see estates, dairies, animal-breeding establishments or the various facilities at Weihen-

54 See 'Auszug aus dem Statut der Koniglichen Bayerischen Akademie in Weihenstephan', cited by Steuert, *Die Koenigliche Bayerische Akademie*, 276–277.

55 *Bericht-Weihenstephan*, 1912–13; Diener, 'Paul Kulisch', 749.

56 *Bericht ueber die T.H. Muenchen* , 1891–92 through 1913–14.

stephan. There is no mention, however, of the Faculty putting on short-courses for farmers, nor of extension work of other kinds. And when in 1908 the Agronomy Institute's acquisition of a new experimental field was announced, it was noted that 'so-called practical experiments are not to be the task of the Institute; its goals will rather consist of establishing the physiological basis of various phenomena and processes in cultivated plants.'[57] As one of the staff put it, Munich had done relatively little in the way of advisory work because the Faculty was 'more scientifically oriented' and 'scientific work requires isolation'.[58]

A particularly striking example of the Munich staff's limited contact with Bavarian farmers is the professor of agricultural chemistry, Franz Soxhlet (1848–1926). Head of the agricultural experiment station attached to the Faculty, Soxhlet had been content to restrict soil-testing at the Station to the laboratory, seeing no need for field-testing, and it was Soxhlet who in the late 1890s had told the Ministry of Education that what the Faculty urgently needed was not a model farm but merely additional lecture-rooms and space for collections.[59] The research for which Soxhlet became known in the late 19th century concerned a method of heating and bottling milk so as to sterilise it, but this achievement seems to have found more recognition from medical organisations than from agricultural ones. Although he occasionally gave lectures to Bavarian farmers' organisations and provided reports to the Bavarian Agricultural Council (*Landwirtschaftsrat in Bayern*), the honorary degree which he received from the University of Halle was in medicine and surgery, and in 1893–94 he received medals from a medical congress in Rome. The following year he became an honorary member of the Medical Association in Munich and in 1909–10, of the Gynecological Society. The only agricultural

57 *Bericht ueber die T.H. Muenchen*, 1907–08, p 24.

58 Respectively, 'mehr rein wissenschaftlich eingestellt' and 'Die wissenschaftliche Arbeit vollzieht sich aber in der Stille' (Ludwig Kiessling, remarks in minutes of 2nd meeting of subcommittee, 13 Oct 1925, BayHStA, MK 19682).

59 Raum, 'Die akademischen Lehrgestalten', 747–8; Raum, 'Nachtraege zu Carl Fraas', 123; 'Kurze Darstellung', attached as appendix 16 to 6 May 1926, Kultusminister to members of Sachverstandigenkommission, BayHStA, MK 19682.

society which elected him to membership was the Royal Swedish Agricultural Academy. Small wonder, then, that some accused him of having ignored the needs of Bavarian farming for a generation.[60]

If extension work was far more developed at Weihenstephan than at Munich, what about the actual research conducted in both institutions? To what extent did it address the problems of Bavarian farming or agricultural industry? In the case of plant-breeding, we have already seen that although work at Weihenstephan was relatively practically-oriented, a comparison with Munich would be meaningless since Carl Kraus, professor of agronomy at Munich from 1901 to 1918, had no adequate experimental field there.[61] But we can obtain a general impression of the orientation of these institutions' research in several fields by surveying the kinds of outlets in which they published their work.

60 *Bericht ueber die T.H. Muenchen*, 1891 to 1914. On the attack, see 10 Feb 1919, unsigned memo to Kultusminister, BayHStA, MK 35953.

61 Since Kraus remained director of the state Plant-Breeding Station at Weihenstephan until 1910, it is possible that was able to retain access to experimental facilities there after his move to Munich, but I have been unable to establish this.

Table 5.2 – Patterns of Publication at Munich and Weihenstephan, 1899–1914[62]

	Scientific	*Technical*[63]	*Popular*[64]	*Other*[65]
Agricultural chemistry:				
Franz Soxhlet (Munich)	0	14	7	0
Ernst Wein (Weih.)	0	13	0	1
Joseph Ahr (Weih.)	0	3	7	1
Animal husbandry:				
Emil Pott (Munich)	0	2	21	3
Ludwig Steuert (Weih.)	0	0	21	10
Carl Kronacher (Weih.)	0	2	6	0

62 The publication details are taken from annual reports of the respective institutions. The disparities in numbers of publications from one individual to another are due not only to publishing activity but also to the fact that some staff were employed at the relevant institution for longer periods than others (eg, Soxhlet was at Munich throughout the period, 1899–1914, while Ahr was at Weihenstephan only from 1910 to 1914.

63 This most common journals in this category were *Vierteljahresschrift des Bayerischen Landwirtschaftsrat, Landwirtschaftliche Versuchstationen, Verhandlungen des Deutschen Landwirtschaftsrats, Zeitschr. fuer das gesamte Brauereiwesen, Mitteilungen des Deutschen Landwirtschafts-Gesellschaft, Agrikultur-Chemische Mitteilungen, Landwirtschaftliches Jahrbuch fuer Bayern, Arbeiten des Deutschen Landwirtschafts-Gesellschaft, Jahrbuch des Deutschen Landwirtschafts-Gesellschaft, Fuehlings Landwirtschaftliche Zeitung.*

64 The most common publications in this category were *Deutsche Landwirtschaftliche Presse, Sueddeutsche Monatshefte, Wochenblatt des Landwirt. Vereins in Bayern, Bayerischer Bauernblatt, Illustrierte Landwirtschaftliche Zeitung, Wiener Landwirtschaftliche Zeitung, Sueddeutsche Landwirtschaftliche Tierzucht, Landwirtschaftliche Hefte, Praktische Blaetter fuer Pflanzenbau und Pflanzenschuetz* as well as a variety of other 'Zeitungen' (newspapers).

65 This category includes mostly books or handbook entries though also a few journals which I have been unable to categorise.

	Scientific	*Technical*[63]	*Popular*[64]	*Other*[65]
Agronomy:				
Martin Wollny (Munich)	0	31	12	5
C. Kraus (Munich '01–14)	0	18	8	2
C. Kraus (Weih. '99–'01)	0	4	32[66]	1
Heinrich Puchner (Weih.)	0	9	-	0
Farm management:				
K. Leisewitz (Munich)	0	1	1	3
A. Schnider (Munich)	0	2	2	0
Friedrich Wagner (Weih.)	0	4	76	3

In Table 5.2 I have categorised the kinds of periodical in which staff at both institutions in agronomy, animal husbandry, farm management and agricultural chemistry published over the period 1899 through 1914, based on information from annual reports. In agricultural chemistry the proportion of publications which were 'popular' vs. 'technical' was little different between the two institutions. Indeed, the pattern of Ernst Wein's publications runs counter to the general orientation at Weihenstephan (though there is a hint that the kind of publications emanating from Weihenstephan might have shifted after Joseph Ahr's arrival in 1910). In animal husbandry, too, the proportions are very similar at each institution. That Emil Pott published overwhelmingly in popular outlets is out of line with the general orientation at Munich, though it is worth noting that all of his 21 popular papers appeared in national or even foreign publications rather than in Bavarian ones. In the case of agronomy and farm management, however, the patterns of publication suggest much closer ties

66 The total of Kraus's and Puchner's 'popular' publications together is 32.

to farming at Weihenstephan than at Munich. In agronomy the contrast between Wollny and Puchner is particularly marked; while the former's publications were skewed strongly toward the technical end, including papers in American and French journals, Puchner's and Kraus's tended the other way, with the former's popular work appearing exclusively in regional outlets. In farm management, similarly, the great majority of Wagner's work was published in popular publications, most of it in a weekly newspaper aimed at Bavaria's small farmers *(Wochenblatt des landwirtschaftlichen Vereins in Bayern)*.[67]

In their teaching, research and extension work, therefore, Munich and Weihenstephan were diametrically opposed in their orientations toward science and practice. Moreover staff at each institution were well aware of this fact and repeatedly emphasised it while appealing for public support during the controversial educational debates of the 1920s, to which we now turn.[68]

A struggle for survival: the attempt to rationalise Bavarian agricultural education during the 1920s

As we saw in Chapter Three, the controversy of 1919–1920 is useful in revealing the different conceptions of agricultural education which divided academies from universities. As a national debate, however, it was conducted at a rather general level with little reference to the

67 Schloegl, *Bayerische Agrargeschichte*, 510–511. Since the *Wochenblatt* was distributed to members of the Landwirtschaftliches Verein (of whom there were 100,000 in 1907), it reached about 20% of the farms in Bavaria (Anon., 'Jahresbericht des Bayerischen Landwirtschaftsrates').

68 For a few examples among many: 22 Nov 1929, minutes of meeting at Kultusministerium, BayHStA, MK 19683; remarks of both Kulisch and Kiessling in minutes of sub-committee meetings on 21 Dec 1925 and 13 Oct 1925, BayHStA, MK 19682; remarks by Ministerialdirektor Matt in minutes of what was probably a committee of the Landtag, n.d. (June 1922?), TUM-Archiv, Sign = I, 8, Bd. 1.

specific political circumstances with which particular institutions had to reckon. Looking more closely at this debate in Bavaria will help to put some meat on those bones. But above all it will be useful in illuminating the model. For the threat of rationalisation during the 1920s brought Weihenstephan and Munich into direct conflict, making visible the process whereby institutions must continually position themselves in relation to agents within both academic and politico-economic fields.[69]

After 1918 the explosion in the number of agriculture students, in Bavaria as elsewhere, meant that increases in expenditure on higher education were going to be necessary.[70] But the weakness of the post-war economy prompted many politicians to ask – yet again – whether Bavaria could afford two agricultural institutions.[71] Thus the long-standing debate over the merits of Munich versus Weihenstephan reemerged, though this time in a particularly aggressive way. Strong opinions – including attempts to defame central figures – were voiced, not just by academics and students at the two institutions, but also by city officials in Freising and Munich, members and committees of the state legislature, professional associations, agricultural journals, and the press.[72] Calls for the closure of one or other institution were frequent, and both Weihenstephan and Munich were continually forced to justify their existence.

For a brief period after the war it looked as though the problem might be resolved. In 1919 staff at Weihenstephan voted unanimously for a solution which their then director (Carl Kraus) had proposed in

69 The existing secondary accounts of this controversy (Raum, 'Die Ereignisse seit 1919' and Diener, 'Paul Kulisch') are quite useful, partly because both authors had first-hand knowledge of the events in question, but also because they drew upon a considerable amount of archival material. But since both authors were employed at Weihenstephan during the 1920s (and obviously remained loyal to it), I have constructed my own account based primarily upon archival and other primary materials.

70 At Munich the number of agriculture students rose between 1913–14 and 1921–22 (both winter-semester) from 186 to 610. At Weihenstephan the increase was from 217 to 508 (Reiner, 'Das bayerische Hochschulwesen', 107).

71 The economic crisis posed a similar threat to some engineering colleges (Gundler, *Technische Bildung*, 339ff).

72 Diener, 'Paul Kulisch'.

the 1890s: a merger of the two institutions with the new hybrid to be based at the engineering college in Munich.[73] Such a move would have given a big boost to Weihenstephan's academic status while simultaneously alleviating the threat of closure facing Munich. Quite why staff at Munich did not jump at the chance is not clear. Evidently they preferred to take their chances as an independent institution rather than merging with a lowly academy, perhaps hoping that the new political climate might reverse their fortunes. In the event, the merger did not go ahead, and Munich's misjudgement was soon to cost them dearly.

Weihenstephan meanwhile was able to command ever more popular support. Late in 1919 the Bavarian Legislature voted (unanimously!) to upgrade the academy to a college, and in 1921 they went further, granting Weihenstephan the right to award the more 'scientific' six-semester degree. This gave Weihenstephan formal parity with Munich in terms of status, thus increasing its threat as a competitor. In addition, the Ministry soon authorised a series of infrastructural improvements at Weihenstephan: new chairs for physics and economics (1922, 1923), an institute for agricultural machinery (1923), extended experimental fields for agricultural chemistry, and greatly expanded facilities for dairy science, physics, chemistry, botany and zoology.[74]

With the gap in facilities widening yet further, rumours again circulated that the Faculty in Munich would be dissolved.[75] In June 1922 things came to a head when the Minister of Education warned the legislature that it would soon need to take a decision on the future of Munich and Weihenstephan. Bavaria could no longer afford to maintain two institutions of higher agricultural education, he argued,

73 6pp appendix to 10 May 1924, Kiessling to Senat der Technischen Hochschule, TUM-Archiv, Sign= I, 8, Bd. 1.

74 *Bayerische Hochschule fuer Landwirtschaft und Brauerei* (1929), 7; Diener, 'Paul Kulisch', 732.

75 14pp typescript of what looks like a transcript of a debate in the Bavarian Landtag, n.d. but probably June 1922, TUM-Archiv, Sign= I, 8, Bd. 1. On rumours: 2 Feb 1921, Vogel to Kultusminister, BayHStA, MK 19697. In a public speech in Dec. 1921 Ludwig Kiessling argued vigourously against closure (Kiessling, 'Die Stellung der Landwirtschaftslehre').

and since Weihenstephan was by far the better equipped of the two, Munich was the natural candidate for closure. Once again most of the legislators who spoke on the issue concurred.[76] Some sections of the agricultural community, nonetheless, spoke out in support of the beleaguered Munich. The Bavarian Chamber of Peasants (*Landesbauernkammer*), for example, argued for the retention and expansion of both institutions, and a petition in defence of the Faculty was published, organised by Jakob Ackermann, an estate-owning member of the legislature who was generally regarded as the leader of big farmers in Bavaria (and who held an honorary doctorate from the Faculty).[77]

In 1923 the debate became especially heated, and Munich continued to lose ground.[78] The following year the Bavarian Legislature once again upgraded Weihenstephan, voting overwhelmingly to grant it the right to award the doctorate. Fearing that this erosion of yet another academic distinction would provide ammunition for their enemies, staff at Munich quickly launched a counterattack, putting a case to the Education Minister that closure (or restructuring) of Weihenstephan would save far more than closure of the Faculty. Moreover, it would make sense, they argued, to turn Weihenstephan back into an academy where it could play an important role. The latter proposal was echoed toward the end of the year by the aforesaid Jakob

76 *Muenchen-Augsburger Abendzeitung*, 27 June 1922; *Bayerische Staatszeitung*, 27 June 1922; remarks by Minister Matt in transcript of a debate in the Landtag (all in TUM-Archiv, Sign= I, 8, Bd. 1).

77 5 Mar 1923, Bayerische Landesbauernkammer to Kultusminister and Min. Landwirtschaft, citing their resolution of 14 Feb 1922 (BayHStA, MK 19682); 'Aufruf zur Erhaltung und zum Ausbau der landwirt. Abteilung der T.H. Muenchen', signed by Ackermann et al. (n.d. but probably Dec 1922, TUM-Archiv, Sign= I, 8, Bd. 1). On Ackermann see Raum 'Die Ereignisse seit 1919', 14; and 30 Jan 1925, Kulisch to Kultusminister, BayHStA, MK 19682. A few years earlier one of Munich's long-term supporters, the Bavarian Agricultural Council, had argued that in view of the very large student demand for agricultural education, they saw no reason for major structural reorganisation of the kind then being contemplated in Prussia (*Wochenblatt des Landwirtschaftlichen Vereins in Bayern*, Vol. 110 (1920), 54.). The Council, however, seems not to have played a prominent role in the controversy of 1922–23.

78 13 May 1924, draft letter, Senat der Technischen Hochschule to Kultusminister, TUM-Archiv, Sign = I, 8, Bd. 1. Cf. Raum 'Die Ereignisse seit 1919', 14.

Ackermann as well as by a provincial newspaper.[79] (This would have meant a return to the status hierarchy of the 1890s where the 'division of labour' between the two institutions would not have posed such a threat to Munich since the two would also occupy different rungs of the ladder.)

Faced with this resistance, the legislature asked the state government in 1925 to establish a committee of experts whose task would be to formulate recommendations for the future shape of higher agricultural education in Bavaria.[80] Duly constituted, the committee chose the soft option. Neither institution was to be closed, but Weihenstephan was to become 'organically linked' to the Faculty in Munich (while in fact maintaining almost complete autonomy). This new loose merger finally came into effect in May 1928,[81] but it was not to last. The German economy continued to falter, and student numbers in Bavaria dropped yet further. Several newspapers protested that this so-called merger was not saving any money as had originally been hoped (voicing a concern which the Finance Ministry had harboured from the beginning). In addition the animosity between the staffs from Weihenstephan and Munich, which had been accumulating since the war, intensified once the two sides were obliged to work together in the new Faculty.[82] New proposals aiming at producing real savings were soon drafted by members of the legislature as well as by staff at the two institutions.[83] By the summer of 1929 the Education Ministry had completed its own proposals, and in the autumn of 1930 a new organisational structure was established. Henceforth students taking the six-semester degree were to spend the first three semesters in

79 13 Nov 1924, J. Ackermann to Ministerpresident Held, Bay HStA, MK 19682; *Moosburger Tagblatt*, 18 Nov 1924, copy in TUM-Archiv, Sign = I, 8 , Bd. 1.

80 27 June 1925, minutes of meeting of Sachverstaendigenkommission; and 2 Sept 1925, Kultusminister to Prof. Dr. Ahr (of Landwirtschaftsministerium), both in BayHStA, MK 19682.

81 30 May 1928, 'Bekanntmachung' issued by the Rektorat der Technischen Hochschule, TUM-Archiv, Sign = I, 8, Bd. 1.

82 *Muenchener Post*, 16–17 Mar 1929; *Der Hopfenbauer*, 15 June 1929; both in BayHStA, MK 19683. Diener, 'Paul Kulisch', 738–9.

83 eg, Raum, 'Kurzgefasste Denkschrift zur landwirt. Hochschulfrage in Bayern', 30 Apr 1929, BayHStA, MK 19683; see also MK 19684.

Munich taking basic sciences and the last three in Weihenstephan taking agricultural sciences as well as their final examinations (a clear case of academic drift, compared to the structure of Weihenstephan's previous curriculum). Those students taking the more practical four-semester course would take all of their courses at Weihenstephan. Understandably Munich's staff were not happy with the new structure because it ceded control over the agricultural sciences largely to Weihenstephan.[84]

Finally in 1933 the Nazis' takeover fundamentally altered the political balance of power, creating new local circumstances which Munich's staff sought to exploit in order to get agricultural education back onto the agenda. Within months of the seizure of power, Ludwig Kiessling was commissioned by the new Bavarian Education Minister, Hans Schemm, to formulate plans for restructuring agricultural education so that *all* teaching would be concentrated at Munich (as Kiessling had proposed in 1929). The language with which Kiessling formulated these plans suggests that he knew how to deploy Nazi rhetoric in order to secure his aims. Moreover he suggested that the opposition to his plan at Weihenstephan was coming from staff members who were simply hostile to National Socialism. This gambit succeeded, perhaps aided by the fact that Himmler had been a student of agriculture at Munich, and in October 1933 Kiessling himself was appointed dean of the reorganised Faculty. From the spring of 1934, therefore, all teaching and examining were conducted in Munich while the facilities at Weihenstephan were used solely for research and practical demonstrations. Unable to reverse the merger, Munich had at least succeeded in rejigging the 1930 arrangements so that they now held the reins. After the war, with Munich badly bombed and the American military government keen to see the resumption of teaching at Weihenstephan, the Faculty moved from Munich to Weihenstephan, therewith realising a plan which Carl Kraus had originally put forward in the 1890s.[85]

84 Raum 'Die Ereignisse seit 1919', 18–22.

85 On rhetoric see 30 Aug 1933, Kiessling to Prof A. Schwaiger, plus enclosure, TUM-Archiv, Sign = I, 28, Bd. 1. On opposition: 8 June 1933, Kiessling to Rektor der Technischen Hochschule, TUM-Archiv, Sign= I, 28, Bd. 1. On

From the establishment of the Faculty in 1872 right through the 1920s, Weihenstephan and Munich had developed markedly different orientations toward science/practice. But which factors were most important in producing this divergence? Let's start with their geographic locations which were so strikingly different. Although Weihenstephan was walking distance from the train station in Freising, the journey to Munich was probably long enough to discourage students from taking courses at the engineering college or the University. And the town of Freising (pop. 15,000 in the 1920s) was not overburdened with cultural attractions or nightspots. Accordingly Weihenstephan was routinely branded by its detractors as an 'isolated' institution. On the other hand, the rural location made it far easier to establish experimental fields or other kinds of demonstration facilities there than in Munich. Geography thus imposed constraints but also offered opportunities to each institution.

But it would be a mistake to see geography as simply a natural constraint upon these institutions for its effects were capable of being mitigated. Moreover, the importance of geography was grossly inflated by both institutions as they constructed images of themselves in an attempt to attract staff and students. For example, Weihenstephan's rural location, with its proximity to various state research and testing stations, allowed it to cultivate the image of being 'close' to agriculture while Munich's students were derided as 'asphalt farmers.' Conversely Munich exploited the presence of its University, theatres and concert-halls to lure students with the prospect of culture, breadth and learning. But in reality Weihenstephan was not so far away from Munich as to cut it off *entirely* from the city's opportunities. From the start, for example, some staff lectured at both Weihenstephan and

Himmler: Raum 'Die Ereignisse seit 1919', 23. On the new structure: 15 Feb 1934, 'Bekanntmachung' from Kultusminister, TUM-Archiv, Sign= I, 28, Bd. 1; 5 Dec 1934, H. Schemm (Kultusminister) to Staatsminister der Finanzen, BayHStA, MK 35953; and Raum, 'Die landwirtschaftliche Hochschullehre'. On post-1945 see Diener, 'Paul Kulisch', 747.

Munich, and as we have seen, Munich's staff often took their students to Weihenstephan for practical demonstrations.[86] Had geography been a serious constraint, it is doubtful whether the merger which occurred in 1933 would have got off the ground. Nor was a big city like Munich necessarily doomed to have poor demonstration facilities; it was instead generally agreed that before the First World War most staff had actually made little *effort* to obtain better facilities. This began to change after the War when the new professors of animal-breeding and agronomy, dissatisfied with the substandard provision, appealed to the Ministry to establish an experimental farm.[87] Had the Ministry wanted to endow Munich with experimental fields on the outskirts of the city, it should have been possible. As late as 1890, for example, the engineering college was still on the northern edge of the city, bordering on farmland, though by the 1920s finance was tight, and investing in Munich did not look like a good bet.[88] Geography, therefore, was in part a rhetorical tool with which Munich and Weihenstephan could jockey for position within the academic and economic fields.

But what about their very different institutional structures? Weihenstephan was after all an independent college while the Faculty in Munich was part of a much larger institution. Unfortunately evidence on this issue is nowhere near as rich as on Breslau, but it does appear that the Faculty's freedom for manoeuver was occasionally

86 Steuert, *Die Koenigliche Bayerische Akademie*, chp. 7. From 1914 Heinrich Puchner, professor of agronomy at Weihenstephan, lectured at Munich (*Bericht ueber die T.H. Muenchen*, 1913–14). During the 1890s a variety of adjunct professors commuted from Munich in order to teach at Weihenstephan (*Die Koenigliche Bayerische Akademie*, 1898).

87 27 July 1929, Kultusminister to Staatsministerium, BayHStA, MK 19683; Henseler, 'Denkschrift fuer den weiteren Ausbau...', Mar 1921, TUM-Archiv, Sign= I, 8, Bd. 1; Raum, 'Die Ereignisse seit 1919', 13. Weihenstephan's rector, Paul Kulisch, was one of those who perceived this demand for facilities as a complete reversal of the Faculty's longstanding position (Diener, 'Paul Kulisch', 735).

88 On the college's location see Raum, 'Nachtraege zu Carl Fraas', 123. On investing in Munich see 14pp transcript of what appear to be minutes of a meeting of the Landtag in late 1921 or early 1922, TUM-Archiv, Sign = I, 8, Bd.1.

restricted. As we saw above, the Engineering College vetoed a proposal in the 1890s that its Agricultural Faculty should merge with forestry at the University, and there are signs that agricultural staff were not made welcome in the College during the 1880s and '90s.[89] Similarly, when proposals for a merger of Weihenstephan and Munich (to create an enlarged Faculty at the College) were discussed in 1925, staff at Weihenstephan were concerned that non-agricultural staff at the College would have too much say in agricultural matters.[90] What is less clear, however, is whether Munich's institutional location had any effect upon its orientation toward science/practice. Its lack of demonstration facilities before the First World War, for example, seems to have been due to attitudes of agricultural staff rather than to pressure from the wider institution. Nor is location in the College apt to have been responsible for the lack of a four-semester practically-oriented degree since such degrees were commonplace at the universities. Furthermore, it is difficult to imagine that the College – an institution devoted after all to the technical sciences – would have harboured anti-utilitarian sympathies comparable to those in the Philosophical Faculties of the universities.[91] While the Munich Faculty's institutional circumstances were sometimes restrictive, therefore, there are no signs that these pushed the Faculty in a 'scientific' direction.

Let's turn, therefore, to the student bodies of the two institutions. In some respects they were quite distinct. Their qualifications, as we have seen, were very different, and they seem to have come with rather different aims (no doubt attracted by the contrasting images of the two institutions). At Weihenstephan many students came to get a qualification; over the period 1895 to 1908 about half of the students

89 Soxhlet, 'Denkschrift'.

90 The proposals evidently allowed for basic science professors at the Engineering College to be members of the committee which would preside over agricultural degrees (document dated 26 Nov 1925, BayHStA, MK 19682). And staff at Weihenstephan were uneasy at the prospect that agricultural staff would be outnumbered in the College's Senate (21 Dec 1925, minutes of meeting of subcommittee, BayHStA, MK 19682).

91 I found no indications in the minutes of the Faculty's appointments committees, for example, of any consistent difference of opinion between the agricultural staff and the representatives of other faculties.

remained for the full two years of the degree (and 30–40% took the final examination). At Munich, however, most students dropped in for a quick taster; about two-thirds of the students left after only a year, and only 5% stayed on for the full three years of the degree.[92] This contrast may have had something to do with socio-economic status, for Munich's students were somewhat more well-to-do than Weihenstephan's and may thus have been less concerned about acquiring a degree. Among those from civil service families, those at Munich were predominantly the sons of top civil servants while those at Weihenstephan were largely the sons of mid-level officials. Similarly, the sons of large farmers were more strongly represented at Munich (30%) than at Weihenstephan (11%). Indeed, along with Goettingen and Jena, Munich had the highest such proportion of all the agricultural institutions while Weihenstephan – along with Breslau and Hohenheim – had the lowest.[93] In other respects, however, the two student bodies were surprisingly similar. In terms of geographical origin, for example, most of the students at both institutions before the First World War were from Bavaria, though the proportion was somewhat higher at Weihenstephan.[94] Given Weihenstephan's practical orientation, one might have expected more of its students to come from farm backgrounds, but in fact fewer of them were actually the sons of farmers (17%) than were Munich's (36%). On balance, therefore, the differences between the two student bodies are not as great as one might have expected, given the marked differences in orientation between Munich and Weihenstephan. This is consistent with the model,

92 Jenne, *Das landwirtschaftliche Unterrichtswesen*, 29 and 54–56; Raum, 'Die Ereignisse seit 1919', 33–37.

93 *Deutsche Hochschulstatistik*, 1928. Other data suggest that while in the late 1920s most agricultural departments and colleges attracted roughly equally numbers of sons of large versus small farmers, at Munich the proportion was heavily skewed toward the former (Lucas, *Die Berufsausbildung*, app. 1; Titze, *Datenhandbuch*, Table 127). Unfortunately I have been unable to find comparative data on social backgrounds for the decade before the First World War.

94 At Munich between 1908 and 1914 about 70% of the students were from Bavaria (*Bericht ueber die T.H. Muenchen*). At Weihenstephan between 1904 and 1914 over 80% were Bavarians (Steuert, *Die Koenigliche Bayerische Akademie*; *Bericht-Weihenstephan*, 1912–13 and 1913–14).

therefore, that as a constituency the student body was not particularly influential.

If the model is correct, however, an institution's place in the academic status hierarchy should be much more important, and this was certainly true in Bavaria. To begin with, because the Faculty was located within a college, it immediately ranked higher on the academic totem pole throughout the pre-1914 era than the 'mere academy' at Weihenstephan. What's more, Munich's students enhanced this status by virtue of the high proportion who possessed the *Abitur*, as did its staff who were more likely than Weihenstephan's to have a doctorate or to have habilitated. And the greater emphasis placed by appointments committees upon 'scientific' qualifications than practical ones ensured that this would continue. Given the fact, however, that Weihenstephan had been endowed with far superior demonstration facilities from the start, there was no way that Munich could hope to compete with Weihenstephan as a 'practical' institution. Munich's strategy for recruiting staff, students and resources, therefore, was to play the status card (just as at Berlin). That Munich should have been the only German agricultural institution during the 1920s which neither offered the four-semester practical degree nor required relevant work-experience of its incoming students thus makes sense. To have done either would have been to embrace the utilitarian values generally associated with institutions at the bottom of the status hierarchy, thus weakening Munich's main claim to legitimacy.

Weihenstephan was much further down the ladder, partly because of its history. It had been only a school until 1895, and it was the last of the agricultural academies to acquire all of the rights of a college.[95] It proved difficult to shake off this pedigree. Even several years after Weihenstephan had become a fully-fledged college, a Munich newspaper claimed that Weihenstephan's doctorates were not yet recognised as a qualification by any German university.[96] The fact

95 It is perhaps significant that among the entries for the agricultural colleges in a major German handbook of higher education, Weihenstephan's was by far the shortest: 11 pp for Berlin, 6 each for Hohenheim and Bonn, and only 2 for Weihenstephan (Lexis, *Die Hochschulen fuer besondere Fachgebiete*).

96 *Muenchener Post*, 16–17 Mar 1929, BayHStA, MK 19683.

that Weihenstephan's staff and students had fewer academic qualifications did not help. A clear indication of the relative status of the two institutions is the fact that between about 1900 and 1920 staff at Weihenstephan sometimes moved to posts at Munich but never, so far as I can tell, the other way around.[97] Given its lowly position but excellent facilities, Weihenstephan also played to its strengths, promoting itself as a practical institution. By doing so it was easier to recruit students who had left school at 16, partly because they had relatively few options within the higher education system, but also because they were often more interested in obtaining vocational qualifications. Thus it made sense to offer just the four-semester degree and to tailor science teaching to agricultural problems; this is what the clientele was looking for. And although this legitimation-strategy did nothing for Weihenstephan's academic status, it did appeal to those sections of the agricultural community – be they students or farmers – who were suspicious of the ivory tower.

Throughout the educational debates of the 1920s the question of academic status was everpresent. And the language in which protagonists cast their arguments clearly indicates that this confrontation was a replay, not only of general 19th century German clashes between competing ideologies of education, but of the specific issues which had divided Julius Kuehn from defenders of the academies in the 1890s. At Munich the conception of education was rooted in the 19th century ideals of *Wissenschaft* and *Bildung*. The curriculum was said to provide a 'deep and well-rounded education'. Its supporters emphasised that students had better opportunities than at Weihenstephan for acquiring 'general cultivation' (*Allgemeinbildung*). Especially important for the agricultural student, Ludwig Kiessling maintained, were history, philosophy and art history. Without the opportunity to attend such courses, students would become mere 'specialists' (*Fachleute*).[98] The closure of Munich's Faculty was un-

97 Among the migrants were Carl Kraus, Theodor Henkel and Ludwig Kiessling.

98 14 Jan 1924, Kiessling to Magnifizenz, TUM-Archiv, Sign= I, 8, Bd. 1. See also 'Aufruf zur Erhaltung und zum Ausbau der Landwirt. Abteilung der T.H. Muenchen', n.d. but probably Dec 1922, TUM-Archiv, Sign = I, 8, Bd. 1. Kiessling, remarks in minutes of meeting of sub-committee on 13 Oct 1925,

thinkable, Kiessling argued, because Bavaria needed 'intellectual leaders and teachers in agriculture who have been educated in the most complete and all-embracing way, just like all other academically educated citizens'.[99]

While invoking this ideal of cultivation to elevate themselves, Munich's spokesmen also deployed it in order to disparage the utilitarian values of their competitors. By insisting that agricultural education should be located in the countryside, Munich maintained, their opponents

> were dragging the debate down from the sphere of academic and higher national and cultural considerations onto the lower level of commercial and party politics, mainly to serve the material interests of certain opportunists [...]. The academic spirit and scholarly viewpoint require higher education to address itself, not to local needs [...] but to universal culture and science.[100]

BayHStA, MK 19682; Kiessling, remarks at 17 Jan 1930 meeting of representatives from both institutions in the Kultusministerium, BayHStA, MK 19684.

99 '[...] einen Nachwuchs an geistigen Fuehrern und Lehrern der Landwirtschaft verlangen, der in der vollkommensten und umfassendsten Weise, ebenso wie alle uebrigen akademischen Buerger, ausgebildet wird.' (5 May 1924, Kiessling to Senat der Technischen Hochschule, TUM-Archiv, Sign= I, 8, Bd. 1).

100 '[...] wurde von den Gegnern der Muenchener Hochschulabteilung aus der Sphaere akademischer und hoeherer staats- und kulturpolitischer Betrachtungsweise auf die niedere Ebene der Geschaefts- und Parteipolitik hauptsaechlich im materiellen Interesse einzelner Nutzniesser herabgezogen [...]. Akademischer Geist und wissenschaftliche Auffassung verlangen nicht einen Hochschulunterricht nach lokalen Beduerfnissen [...] sondern im universellen Sinne der Kultur und Wissenschaft.' (21 Dec 1929, Senat der Technischen Hochschule to Kultusminister, BayHStA, MK 19683). In their bid to attract students, members of the Faculty deployed the same rhetoric: 'It is absolutely essential that future civil servants [...] as well as those who will run farms partake, not only of technical knowledge, but of the nation's general cultural heritage as well.' (Henseler, 'Die landwirt. Abteilung der T.H. Muenchen', 411).

The proposals for reorganisation emanating from Weihenstephan, they complained, were not appropriate for a college or university but only for a technical school (*Fachschule*).[101]

These imputations of inferiority, of course, were sharply contested by Weihenstephan's rector. A science concerned with application, Paul Kulisch objected, was most definitely not a second class science but rather one which embraced other goals and methods. Indeed, in many respects such a science was more demanding than a pure science.[102] The importance so often attached to 'general cultivation', he urged, ought not to be exaggerated. If a student lost his way while dabbling in every imaginable subject while mastering none, his education did not thereby become 'superior' (*hochstehend*). It was far preferable to take a narrower range of subjects and study them in depth. '[...] ties to the *Universitas litterarum*', as he put it, 'should not be pursued if the cost of doing so is to weaken the connections between agricultural research and the economy.'[103] Here was a diametrically opposed conception of agricultural education and one likely to appeal to a very different audience. Ironically, the arguments which Munich's staff chose to deploy against Weihenstephan were precisely those which had been used against the engineering colleges themselves (namely by the universities from the 1870s): the inferiority of

101 22 Nov 1929, remarks by the Rector of T.H. Muenchen in minutes of a meeting at Kultusministerium, BayHStA, MK 19683. Following the merger of the two institutions in 1930, Munich's students were faced with the prospect of having to transfer to Weihenstephan for their final three semesters. The students' association's response was to recommend that students transfer instead to a university-institute (outside Bavaria) for these semesters 'in the interests of social status' (*des Standesansehens*)(Raum, 'Die Ereignisse seit 1919', 21).

102 Kulisch, remarks to meeting of a sub-committee on 13 Oct 1925; cf his remarks to an earlier meeting of the committee on 21 Dec 1925, both in BayHStA, MK 19682.

103 Reprint of Kulisch, 'Die Bayerische Hochschule fuer Landwirtschaft und Brauerei in Weihenstephan', *Die Landwirtschaft in Bayern* (ca. May/June 1929), p.104 (BayHStA, MK 19683). On the advantages of depth see Kulisch's remarks in 'Niederschrift ueber die 2. Sitzung des Unterausschusses', 13 Oct 1925, p. 19, BayHStA, MK 19682. In a similar vein see Raum, 'Kurzgefasste Denkschrift zur landwirt. Hochschulfrage in Bayern', 30 Apr 1929, BayHStA, MK 19683.

specialisation to breadth and of utility to purity. But now the shoe was on the other foot. Munich was having to appropriate the arguments of its former enemies in order to prevent a practically oriented newcomer from gaining further ground.

According to the model, however, the strategies which staff adopt in order to advance their institution's development should only prevail when they enjoy state sanction. And this is just what happened in Bavaria. From the 1890s through the 1920s, as I hope this chapter has made clear, the Ministry chose consistently to bestow ever more lavish facilities upon Weihenstephan while ignoring such requests (especially after the First World War) from Munich. When student numbers at Munich were low in the late 19th century and pressure to close the Faculty grew, or when Weihenstephan was upgraded to an academy in 1895, the Ministry's response was to attempt to reduce competition between the two by emphasising their different roles: Munich was to be seen as a 'scientific' institution while Weihenstephan was a 'practical' one. Where necessary the Minister was prepared to override staff reservations about candidates' academic qualifications in order to ensure that at Weihenstephan those with strong practical backgrounds were appointed (eg, Anton Fehr). And in 1930 it was the Minister who finally pushed through a merger of the two institutions (in a form which more nearly met the requests of staff at Weihenstephan than those at Munich).

In doing so, however, the Minister was to some extent responding to political pressure, whether from the state legislature or from particular interest groups. And this brings us finally to the politico-economic field. Since Bavaria's economy was predominantly agricultural, one might expect the players within this field to have had a substantial impact upon higher agricultural education. Given the very different strategies with which Munich and Weihenstephan had historically sought to justify their existence, however, such an impact was likely to favour Weihenstephan. During the 1920s debates, accordingly, Munich sought to minimise the significance of economic considerations. In 1925, for example, a discussion took place as to the appropriate membership of the expert-committee which was to make recommendations for the reorganisation of agricultural education. The Education Minister suggested that the committee should include representatives from the farming community (*der praktischen Landwirtschaft*) and from the brewing industry as well as a variety of government officials and a representative from the University. But the spokesman for Munich, Ludwig Kiessling, demurred, arguing that the issue was a 'purely academic' one (*reine Hochschulfrage*). It should, therefore, be resolved by academics alone. 'All other persons, including farmers, lacked the requisite expertise to pronounce on the matter.' Needless to say, Weihenstephan's rector, Paul Kulisch, objected that the issue was not purely academic since agricultural education was a matter of direct concern to farmers.[104]

Knowing the name of the game, however, Munich sought at the same time to recruit whatever support it could from particular sections of the agricultural community. Since the students who took its six-semester degree were qualified for agricultural-administrative jobs in either the civil service or the private sector, two professional associations representing these occupations (*Verband der Landwirtschaftsberater Bayerns* and the *Reichsbund akademisch gebildeter Landwirte in Bayern*) were a likely possibility. Another came from estate-owners. As Table 2.1 indicates, although Bavaria's farms were overwhelmingly small, about 0.1% of them (amounting to about 500 farms) were estates of 100 hectars or more. That provided a further potential constituency which could be tapped, along with the fact that nearly a third of Munich's students were the sons of estate-owners. This pattern of support is evident from Ackermann's 1922 petition backing the Faculty. Of the 84 signatories, there were 32 estate-owners (*Gutsbesitzer* or *-paechter*), 13 high-level government officials, 11 agricultural administrators (either directors of agricultural secondary schools or administrators of other organisations), 6 bankers, 7 professors, 8 individuals listed as having an academic degree or an honorific title (*Hofrat, Landwirtschaftlicher-Rat, Oekonomie-Rat*), and two who simply listed their occupation as 'farmer'.[105] The same

104 27 June 1925, minutes of meeting of Sachverstaendigenkommission, BayHStA, MK 19682.

105 'Aufruf zur Erhaltung und zum Ausbau der landwirt. Abteilung der T.H. Muenchen', signed by Ackermann et al. (n.d. but probably Dec 1922), TUM-Archiv, Sign= I, 8, Bd. 1. The pattern of Munich's support was recognised by

pattern is also noticeable in the honorary doctorates which Munich awarded. Whereas Weihenstephan awarded most of its honorary degrees during the 1920s to politicians and officials in agriculture and education, Munich awarded most of its degrees to estate-owners.[106]

That Weihenstephan should have honored so many politicians and top civil servants reflected the fact that it could count upon a wider political base than could Munich. To see why, we need to consider briefly the changing complexion of agricultural politics in Bavaria since the late 19th century. Before the 1890s peasant participation in politics at state or Reich level had been minimal. The only organisation claiming to represent their interests had been the Agricultural Association of Bavaria (*Landwirtschaftlicher Verein in Bayern*, est. 1810), though since membership was voluntary, it never enrolled more than 20% of the peasantry. Moreover, its leadership up to the First World War was dominated by nobles and an intelligentsia consisting of officials, academics and priests. The structure of this leadership, along with the fact that the Association had been set up with government support as an explicitly non-political association in order to promote technical improvements in peasant agriculture and that it was reliant upon state subsidy, meant that it was in no position to challenge state agricultural policy. As a result, its worthy but

both sides: see 21 Dec 1929, Senat der Technischen Hochschule to Kultusminister, BayHStA, MK 19683; Kulisch, 'Gesichtspunkte zur Entscheidung ueber den Ausbau des landwirt. Hochschul-Unterrichtes in Bayern', 17 Jan 1923; and Kulisch to Kultusminister, 30 Jan 1925, both in BayHStA, MK 19682.

106 The five honorary doctorates which were awarded by Weihenstephan in 1927 went to an estate-owner, three civil servants (in Reich Agricultural Ministry, the Bavarian Education Ministry, and the Bavarian Agriculture Ministry), and the representative for the town of Freising in the state legislature. By contrast the five such doctorates which the Engineering College awarded during the 1920s went to an industrialist, three estate-owners, and one titled individual who was honoured for his contributions to agricultural technology (Raum, 'Die Doktorpromotionen', 43–44). The five honorary degrees which Munich bestowed between 1905 and 1918 also went mainly to the social elite rather than to political functionaries. There were two titled estate-owners and ex-presidents of the Bavarian Agricultural Council, one academic, one titled military officer, and one titled civil servant (Dienel and Hilz, *125 Jahre T.U. Muenchen*, 61–62).

patronising attempts to elevate peasant agriculture through continuing education, prize-giving, and at times incomprehensible public lectures, coupled with its inability to articulate peasants' growing economic and social grievances during the troubled 1890s, came to be resented.[107]

Until the 1890s the dominant party in Bavaria, the Catholic Centre Party, had largely ignored peasant dissatisfaction. Reacting against Bismarck's campaign in the 1870s against 'Catholic influence' in the new Germany, the Centre had concentrated on defending confessional interests. In the early 1890s, however, a sharp decline in grain and meat prices, precipitated by a reduction in protective tariffs as well as a drop on the world market, triggered an economic crisis and growing politicisation among farmers, large and small. An important consequence was the formation of a secular (and often anti-clerical) radical alternative to the Centre Party, the Bavarian Peasants League (*Bayerischer Bauernbund*). Responding to the peasants' grievances, the League took large numbers of votes away from the Centre in both national and regional elections during the 1890s.[108] The Peasants League called for the restoration of protective tariffs, a more progressive tax structure which would no longer favour estate-owners, the breakup of large estates, greater access by peasants to the forests on large estates for hunting and wood-gathering, and the abolition of a special land tax which had been imposed on peasants during land-reform in 1848 in order to compensate landlords for the loss of traditional privileges. The democratic thrust of the League's politics was also evident in their call for universal male suffrage and the abolition of the unelected upper house of the Reichstag. Similarly, the League was also determined from the start to replace peasants' longstanding political reliance upon the nobles, officials and priests of the Agricultural Association with a new leadership of educated peasants.

107 Farr, 'Peasant protest in the Empire'; Schloegl, *Bayerische Agrargeschichte*, 556–560.

108 Farr, 'Populism in the countryside'; Hochberger, *Der Bayerische Bauernbund*; Bergmann, *Der Bayerische Bauernbund*; Fritsch and Gottwald, 'Bayerische Bauernbund'.

The Centre Party in Bavaria was quick to get the message. In order to build a relationship with its peasant constituents, it promoted the formation of a network of local peasants associations between 1893 and 1897 which amalgamated in 1898 to form a national organisation, the Christian Peasants Association (*Christlicher Bauernverein*) which was affiliated to the Centre Party. This Association proved extremely successful, partly because it largely appropriated the political platform of the Peasants League but was better organised, and its promotion of self-help via cooperatives was popular. The dominant figure within the Christian Peasants Association before the First World War was Georg Heim, who helped to unite the local associations, prompted them to provide a range of financial and legal services for their members, and established a school specifically designed to train peasant leaders for each locality.[109] This rapid rise of peasant radicalism in the 1890s put agricultural issues at the forefront of Bavarian politics for the next decade, and the government responded with a series of inquiries into the state of agriculture as well as practical measures (eg, a state-funded agricultural bank, assistance to cooperatives, and credits in aid of technical improvements).[110]

In terms of the politico-economic field, therefore, there was a noticeable shift of power during this period from the rural elites of the Agricultural Association to the peasantry, and this continued after 1918. After the War the Peasants League joined with the social-democrats in the short-lived revolutionary soviet government of 1918–19. When the Bavarian Ministry of Agriculture was formed in 1919, accordingly, its first minister was drawn from the Peasants League, as were subsequent ones through 1930.[111] The Centre Party in Bavaria, understandably hostile to the new socialist government, sought to regroup in order to block its influence. In 1918, therefore, Georg Heim and others broke away from the Centre Party to found the Bavarian Peoples Party, a centre-right party representing Catholic peasants and workers which was to dominate all Bavarian governments for the rest

109 Haushofer, 'Baeuerliche Fuehrungsschichten'.

110 Farr, 'Peasant protest in the Empire'; Hochberger, *Der Bayerische Bauernbund.*

111 Haushofer, *Ein halbes Jahrhundert.*

of the Weimar period.[112] In addition to his leading role in this party and in the Christian Peasants Association, Heim became first chairman of the newly formed Bavarian Chamber of Peasants (*Landesbauernkammer*)[113] and was thus arguably the most important figure in interwar peasant politics.

At this point we need to ask how the politico-economic field, thus constituted, affected the development of agricultural education. Upon whom, for example, could the two Bavarian institutions call for support? The Faculty at Munich had particularly close ties to the Agricultural Association. To begin with, it was a campaign by the Association which had led to the establishment of the Faculty in 1872. Moreover the Faculty continued to enjoy the support of the Association thereafter: two of the five honorary doctorates which it awarded between 1905 and 1918 went to past presidents of the Bavarian Agricultural Council (*Bayerischer Landwirtschaftsrat*) which was the highest level committee within the Agricultural Association. The Council served as the Bavarian government's advisory body on agricultural matters and had called several times for the expansion of the Faculty.[114] And as we have seen, the Faculty's main support during the educational battles of the 1920s came from the same two factions which had dominated the Agricultural Association right up to the First World War: estate-owners and officials.

In contrast, the history of Weihenstephan reveals numerous links to both wings of the peasants' movement. That the Munich Faculty's call for Weihenstephan to be closed in the early 1890s was ignored,

112 Bergmann, *Der Bayerische Bauernbund*, 46–49.

113 Schloegl, *Bayerische Agrargeschichte*; Bergmann, *Der Bayerische Bauernbund*. Established in 1920 as the official representative of peasant interests, the Peasants' Chamber thus freed peasants from their traditional dependence upon the Bavarian Agricultural Association as spokesman.

114 'Aufruf zur Erhaltung und zum Ausbau der landwirt. Abteilung der T.H. Muenchen', signed by Ackermann et al. (n.d. but probably Dec 1922), TUM-Archiv, Sign= I, 8, Bd. 1. The leader of this campaign, Jakob Ackermann, was a member of the Council (Bergmann, *Der Bayerische Bauernbund*, 378). In other regions of Germany, too, members of the local Agricultural Association often focussed on higher education while peasants were more concerned with agricultural schooling (Blackbourn, 'Peasants and politics', 55–56).

and that instead Weihenstephan was upgraded in 1895 and received a good deal more investment over the following decade than did Munich, may have owed more than a little to the Bavarian government's unwillingness to provoke peasant leaders at that time since Weihenstephan was widely perceived to be the more accessible and practical of the two institutions. One of the Peasant League's first demands in the Bavarian parliament had been for a stronger practical orientation in agricultural education, and during the 1920s some observers reckoned Weihenstephan to have had strong support from both the Peasants League and the Bavarian Peoples' Party.[115] That does seem to have been the case. One of the four honorary presidents of Weihenstephan's supporters' organisation throughout the 1920s, for example, was none other than Georg Heim. Moreover Anton Fehr (professor of dairy science at Weihenstephan from 1917) was a major figure in post-war agricultural politics. As leader of the right wing of the Bavarian Peasants League after the War, he represented the party in the Reichstag from 1917 to 1933, served as Bavarian Minister of Agriculture from 1924 to 1930, and was Reich Minister of Agriculture for a brief period in 1922. It was through his efforts that a major state dairy research station (Sueddeutsche Forschungsanstalt fuer Milchwirtschaft) was established at Weihenstephan in 1923 (which may be one reason why he collected an honorary doctorate from Weihenstephan four years later).[116]

115 Hochberger, *Der Bayerische Bauernbund*, 254; on the 1920s see Diener, 'Paul Kulisch', 739. Hans Rauch, professor of agricultural accounting and commerce since 1917, represented the Bavarian Peoples Party in the state parliament intermittently from 1917 to 1924 (Bergmann, *Der Bayerische Bauernbund*, 385–386). It may seem odd that Weihenstephan could attract support from two parties which were mutually hostile during the 1920s, but the Peasants League was a broad church, and its right wing was quite close to the Bavarian Peoples Party (Bergmann, *Der Bayerische Bauernbund*, 275, 333–335).

116 Raum, 'Die von 1951–1972 verstorbenen'; Schloegl, *Bayerische Agrargeschichte*, 569; Haushofer, *Ein halbes Jahrhundert*, 28–31; Bergmann, *Der Bayerische Bauernbund*, 22 and passim. Weihenstephan's links to the Bavarian Agriculture Ministry were probably also strengthened by the fact that Josef Ahr, professor of agricultural chemistry at Weihenstephan from 1910 to 1919, was a top civil servant in the Ministry from 1919 through 1928 (Raum, 'Die akademischen Lehrgestalten', 753; Schloegl, *Bayerische Agrargeschichte*, 855).

That Weihenstephan should have enjoyed such abundant political support is hardly surprising, for the College made a concerted attempt to address the needs of peasants. Although very few of Weihenstephan's students were actually the sons of peasants,[117] much of the college's extension work was either accessible to, or explicitly directed at, peasants. Some laboratory tests, for example, were provided at no cost, as was management advice (when solicited by peasants), and some of the accounting courses were tailored for peasants.[118] The Institute of Animal Husbandry devoted a good deal of time to extension work, and in 1912–13 its short-courses appear to have attracted more sons of peasants than sons of estate-owners. In 1923 Weihenstephan began to publish a series of pamphlets aimed specifically at peasants; written by academics, agricultural advisors and farmers, the series is said to have been a great success. In any event, the frequency of these and other contacts with the farming community was often emphasised in Weihenstephan's annual reports.[119]

Throughout the 1920s Weihenstephan's rector, Paul Kulisch, repeatedly stressed the importance of addressing the needs of peasants. He pointed with pride at the large numbers of farmers who came to Weihenstephan each year for continuing education and was keen to expand this activity. At his college, he maintained, there had always been a 'tight connection between theory and practice'; staff had to be *au fait* not only with the relevant technology but with economic conditions as well. In addition, of course, Kulisch used Weihenstephan's peasant-orientation to beat Munich wherever he could. In a region like Bavaria where farms were small, he noted, higher agricultural education had to address the needs, not of the handful of large estate-owners near Munich, but of the small farmer. The oft-cited difficulty in transferring technological advances to the farmer, he argued,

117 According to *Deutsche Hochschulstatistik* (1928), only 6% of Weihenstephan's students' fathers owned small- or medium-sized farms.

118 *Bericht-Weihenstephan* for 1913–14 and 1918–19; *Bayerische Hochschule fuer Landwirtschaft und Brauerei* (1929), 17.

119 On animal husbandry see pp 175–190, file 33, box 1, C.B. Hutchison papers. On the pamphlet series ('Weihenstephaner Schriftensammlung') see Schloegl, *Bayerische Agrargeschichte*.

was due in part to the failure of educational institutions to recognise the peculiar needs and limitations of the small farm, and Munich's arms-length relationship to the farming community merely aggravated this problem. This seems not to have been just opportunistic posturing on Kulisch's part, for his concern with the needs of small farms went back to his days as director of the Agricultural Experiment Station at Colmar in Alsace from 1900 to 1918. A major function of the Station, like its counterpart at Weihenstephan, had been to improve the quality of seed used on the predominantly small Alsatian farms. When the Station discovered that imported high-yielding varieties performed more poorly on local farms than did improved local varieties, Kulisch warned Alsatian farmers to be wary of the elaborate claims made for expensive high-bred seed.[120] In Weihenstephan's case, strong support from peasant farmers meant that the College could count on the public purse being generous.[121]

120 Kulisch, 'Gesichtspunkte zur Entscheidung ueber den Ausbau des landwirt. Hochschul-Unterrichtes in Bayern', 17 Jan 1923; and Kulisch, remarks in minutes of a meeting of sub-committee, 13 Oct 1925; and Kulisch, remarks to 29 Sept 1925 meeting of sub-committee, all in BayHStA, MK 19682. 14 May 1931, Kulisch to Kultusminister/Ministerialrat, BayHStA, MK 19684. On Alsace see Kulisch, 'Die staatliche Foerderung' and Anon., 'Ueber die Zuechtungsversuche'.

121 The College also sought funding from the private sector. In 1923 an 'Association of the Friends and Supporters of Weihenstephan' (*Weihenstephaner Hochschulgesellschaft*) was established in order to raise money for teaching and research. To judge from the lists of officers, donors and members, however, the largest economic sector represented was the brewing industry, whose interests were evidently directed at Weihenstephan's famous school of brewing rather than at agricultural education or research ('Verzeichnis der Vorstandsmitglieder der Weihenstephaner Hochschulgesellschaft', UA-Weih, Sekt II, Fach 71, Nr. 1; *Bayerische Hochschule fuer Landwirtschaft und Brauerei*, 1925; 'Mitgliederliste der Weihenstephaner Hochschulgesellschaft', typescript, n.d., BayHStA, MK 40145.

Conclusion

That Weihenstephan became such a well-endowed and practically oriented institution, therefore, owed a great deal to the structure of the politico-economic field in Bavaria at that time, but also to its location within the academic field. Though staff there no doubt smarted at their lowly position within the status hierarchy, there was no realistic way in which Weihenstephan could hope to use its status to attract staff, students or investment, given that Munich held a far stronger hand in that respect. But as an institution which had once been a school, was attached to a state farm, and had been prodded in a practical direction by the Ministry, Weihenstephan had an advantage from the outset in appealing for support from peasants' organisations who probably perceived it, not as a lofty and inaccessible ivory tower, but as an altogether down-to-earth place from which they might actually benefit. By pursuing a 'practical' strategy, therefore, Weihenstephan was able not only to grow, but even to enhance its formal academic status several times between 1895 and 1924. Faced with an increasingly militant peasantry before the First World War and a politically powerful one afterward, the Ministry of Education could hardly fail to oblige.

The Faculty at Munich, on the other hand, occupied from the start a different locus within the academic field. As one of the wave of new Liebig-inspired institutions then being founded all over Germany, it started life as a place in which the sciences were to play a larger role than farm management or the other agricultural disciplines. Moreover the structure of the politico-economic field at that time was also conducive to a 'scientific' orientation, for the only quasi-official body then representing the farming community was the Agricultural Association. Dominated by estate-owners and officials, the Association looked benevolently upon a scientifically-oriented and high-status educational institution. Nor was there much political pressure for Munich to develop in a more practical direction during the 1870s and 1880s since the government was then relatively unconcerned with agricultural issues. As unrest emerged in the countryside during the

1890s, however, the political-economic field shifted markedly, but the government's response effectively locked Munich ever more tightly onto a 'scientific' path. By declaring a division of labour between the two institutions, the Ministry was able to protect Munich against calls for closure, but with Weihenstephan henceforth designated as Bavaria's 'practical' institution, Munich was destined as the 'scientific' one to remain less well funded (aided by the fact that its staff at that time displayed little interest in shifting in a more practical direction). After the War, to be sure, some of Munich's new appointees tried to break out of the Faculty's historical trajectory by calling for expensive new demonstration facilities. But by then the political-economic field had changed yet again. With the peasantry strong in the legislature, the Agricultural Association marginalised, and public spending curtailed, any future investment was bound to go to Weihenstephan. Munich had by now lost all room for manoeuver. Stuck with its legacy as a 'scientific' institution and a political climate in which that legacy lacked all credibility, the best it could hope for was to avoid closure.

The Bavarian case illustrates several important features of the 'field' concept outlined in Chapter Two. First, a field presents both opportunities and constraints to the institutions occupying it. Placed high in the status hierarchy, for example, Munich was able to attract staff and students who were well-qualified academically though less so in terms of practical experience, and the converse applied at Weihenstephan. Turning to the politico-economic field, we have seen that it was constituted by a *variety* of agricultural interest groups; and from the point of view of a given educational institution, some of these were easier to cultivate as allies than others. Moreover this field was a dynamic one since the power-relations among these groups changed over time.

A second feature of the field is that 'history matters' (or 'chickens come home to roost'), deriving from the simple fact that the actors within a field have memories. When interest-groups were considering whether to establish links with a particular institution, the latter's historical trajectory of actions and attitudes was taken into account since these provided some indication of its future aims. After the First World War, for example, even though some of Munich's new staff were looking more favourably upon 'practice', there was little prospect of peasant organisations turning a blind eye to Munich's past or lending their support for its future.

Finally, in working out a strategy for development each institution needed to take into account, not only the shifting array of interest-groups from whom support might be sought, but also the positions within the academic field that were occupied by competing institutions. For example, so long as Weihenstephan was merely a school, Munich probably enjoyed a certain amount of latitude in the strategy of development which its staff could pursue. As Bavaria's only higher educational institution at that time, it might have opted to move in a 'scientific' direction like Halle or in a practical one like Breslau. But once Weihenstephan had entered the ranks of higher education (1895) and especially after being formally constituted as a college (1919), Munich's freedom of manoeuver was seriously limited. Faced with a competing institution of the same *kind* which was also better equipped, Munich was then under much more pressure to differentiate itself by adopting an orientation toward science/practice which was sharply distinct from that of its competitor.

In deciding how to design the curriculum, which problems are worthy of research, or whom to appoint to a chair, therefore, the members of institutions had to size up the possibilities presented by several fields, all of which were in flux, while keeping an eye on simultaneous moves being undertaken by competitors *and* reckoning with the historical trajectories of all actors. In this way, I think, the field concept is rather good at capturing the general point that human beings 'make their own history though not necessarily as they choose'.

Chapter 6: Conclusion

Having looked comparatively at a number of agricultural institutions and followed the development of two of them over a longer period, we now have a model of institutional dynamics. Its basic principles are these. If we focus on any given institution, staff were almost always concerned to enhance their institution's academic status, a move whose consequence was generally to strengthen its science-orientation. The amount of effort they devoted to that cause varied, however, depending on the institution's rank on the status hierarchy, whether agriculture was situated in a Philosophical Faculty, and its geographic location (since developing a stronger science-orientation was easier at urban institutions than at rural ones). Whether staff succeeded in such campaigns, however, depended upon the stances adopted by both the ministry concerned and interest-groups in the local agricultural economy. If the minister was largely indifferent to the orientation of the institution, he might honour staff's request by upgrading it, or where upgrading was not at issue, he might simply allow the institution to persist largely decoupled from the economy, even for several decades. In other cases the minister might veto staff requests or indeed try to introduce a stronger practical orientation at an institution, especially if under pressure from agricultural interest-groups. Formal political channels, however, were not the only route through which the latter could exert influence. By targetting funding for particular initiatives at a local college, firms or growers associations could also ensure that its research or teaching met their needs. This, then, is the model, but how general is it? In the sections below I show how the model might prove helpful in understanding three different sets of problem.

Extending the model to agricultural education elsewhere

When one turns to other countries after having studied Germany, the similarities are striking: the kinds of institutions in which agriculture was taught, the contrast between 'science'- and 'practice-oriented' institutions, the trend toward academisation, and the prominent roles of the state and of interest-groups in this process. And even where the circumstances in these countries were distinct from those in Germany, the differences between them can nonetheless be conceptualised in terms of the interaction between academic and politico-economic fields. Although little attention has so far been devoted to explaining these phenomena in other national contexts, I think it is now clearer how this might be done.

Britain lends itself particularly nicely to the model. As Stewart Richards has pointed out, agricultural education there was also caught between the conflicting suspicions of farmers and academics in the basic sciences. As in Germany, agriculture was taught both in university departments and in independent colleges of agriculture, with the curricula in the former tending to be more science-oriented.[1] At most universities, for example, students took basic sciences in the first year (often together with science students), encountering agricultural subjects and practical work in agriculture only in their second and final years. Since instruction took place almost entirely in the lecture hall and laboratory rather than in the field, students acquired very little familiarity with practical agriculture. During the interwar years, accordingly, Leeds was the only university to require that students should have spent a year working on a farm prior to entry.[2] The orientation at Oxford was similar; in 1891 a committee charged with setting up an agricultural degree there concluded that suitable lectures

1 Stewart Richards, *Science, Agriculture and the Universities*, chp. 3; Richards, 'Masters of arts'. A notable exception was Wye College whose Oxford- and Cambridge-educated staff gave the curriculum a heavy science-orientation (Richards, 'South-Eastern Agricultural College').

2 Richards, *Science, Agriculture and the Universities*, chp. 4. A similar process appears to have affected Dutch agricultural education (Maat, *Science Cultivating Practice*, chp. 3).

were already available in botany, chemistry, physics and physiology and that agricultural subjects – taught by the sole professor of agriculture – could simply be grafted on. Although that particular proposal was rejected, the degree eventually approved was designed to give gentleman farmers a liberal education, focussing not upon sciences but upon the history, economics, law and organisation of agriculture.[3]

On the other hand, by no means all British institutions between the wars were so science-oriented. A government report on veterinary education in the 1930s, for example, complaining that veterinary education was far too practically-oriented, reads like a critique of late 19th century German academies. Students at the veterinary colleges, it stated, were getting a 'narrow vocational training', and their science instruction lacked the breadth of university science. Veterinary students should therefore take degrees in veterinary science rather than just a diploma in veterinary medicine. There was no research spirit at the colleges, the report continued, because staff were isolated from relevant sciences at the universities. College staff should be allowed time for research which was 'fundamental and disinterested [... and] undertaken for the advancement of basic knowledge'.[4]

As far as the agricultural sciences in Britain are concerned, historical studies of particular disciplines are still rare. But Paolo Palladino's studies of early 20th century plant-breeding have revealed two quite distinct assessments of the importance of genetic theory among academic breeders which seem to have been associated with differing institutional contexts and which correspond to the science/practice orientations discussed above.[5] At Cambridge University's School of Agriculture R.H. Biffen was convinced that breeding practice could only be improved by applying and further developing the Mendelian theory. And in keeping with the dominant view among his

3 Richards, *Science, Agriculture and the Universities*, chp. 3; Morrell, *Science at Oxford*, 113–115.

4 Min. Agriculture, *Report of the Committee on Veterinary Education*, pp 18 and 88, respectively. I am indebted to Richard Hankins for this and other references on veterinary education.

5 Palladino, 'Between craft and science'.

colleagues at the School, he felt that utilitarian considerations need play no part in guiding research in agricultural science. At University College Reading by contrast, John Percival was altogether more sceptical. Though agreeing that genetic theory offered a better understanding of breeding phenomena, he felt it was of limited applicability to those traits of most interest to farmers as well as to actual growing conditions in the field, and his advice to farmers took into account the economic pressures under which they were working. Unlike Biffen, Percival maintained an active interest in agricultural extension, and the orientation of the curriculum at Reading was substantially more practical than at Cambridge.

A similar bifurcation was to be found in the pattern of early 20th century funding of agricultural sciences. The earliest of Britain's research councils, the Development Commission (est. 1910), was particularly keen to fund the development of basic sciences relevant to agriculture, and genetics, plant and animal physiology, and bacteriology benefitted accordingly. The Ministry of Agriculture, however, expressed serious reservations about this policy, taking the view that the needs of farming rather than the judgments of scientists should inform the research agenda, but by the First World War the Development Commission had largely got its way. During the interwar period, for example, the Development Commission favoured the 'fundamental' work done at the Rothamsted Experimental Station over the more practically-oriented work at the National Institute of Dairy Research.[6] The Commission's successor, the Agricultural Research Council (est. 1931), showed a similar preference for fundamental research (which was not surprising, given the elite scientists who sat on both the Commission and the ARC).

Another country whose agricultural education lends itself to the distinction between 'science' and 'practice' is the Netherlands. Harro Maat's history of the agricultural college at Wageningen, 1863–1986, reveals a number of similarities with the German scene at that time.[7] Like Weihenstephan and Hohenheim, Wageningen underwent a simi-

6 Olby, 'The Development Commission'; DeJager, 'Pure science and practical interests'; Vernon, 'Science for the farmer?'.

7 Maat, *Science Cultivating Practice*, chps. 4 and 8.

lar sequence of stages via upgrading: established as an agricultural secondary school in 1876 (and acquiring a model farm a few years later), it became an agricultural college in 1918, an 'agricultural university' in 1986 and finally a mainstream 'university' in the late 1990s. Given these modest origins, it is not surprising that, as in Germany, basic scientists in the Netherlands often regarded agricultural science as a second-rate subject in need of more academic rigour. And just as in Germany, upgrading entailed a shift in the curriculum. Although throughout its history educational reformers at Wageningen periodically advanced either a broad, integrated and more practice-oriented approach or a narrower science-oriented one, over time the latter came to prevail. When formal linkages between Wageningen and applied research institutes as well as experiment stations were discontinued, similarly, research at the College gradually drifted away from the world of practice. What then were the moving forces behind academisation at Wageningen? Since the College was the only Dutch institution teaching scientific agriculture at that time, Maat was probably hesitant to generalise from his case study, but his findings appear to be broadly consistent with the model of institutional dynamics I have developed here. For he sees academic staff as having generally favoured the overall academising trend and to have been encouraged in this respect by the Ministry of Agriculture (at least until the 1980s). Which interest groups resisted this trend is less clear but evidently included agricultural societies (in the late 19th century), former graduates (during the 1930s), and radical students (in the late 1960s).

Turning to the United States, one might have expected that agricultural scientists there were subject to somewhat different pressures than in Europe. Many years ago, for example, Richard Hofstadter argued that various features of the new American republic created a climate in which ivory tower scholarship commanded less respect – and practical common sense more – than in Europe.[8] This would seem to imply that American scholars in the practical sciences were less

8 Hofstadter, *Anti-Intellectualism in American Life*. 19th century travellers to the United States often had the impression that there was less prejudice against manual labour there than elsewhere (Kranakis, *Constructing a Bridge*, 234–235).

subject to condescension from their peers in basic disciplines; if so, they might have been less preoccupied with questions of status so that academic drift was less likely. This thesis finds support from the impressions gathered by various agricultural scientists, both German and American, who crossed the Atlantic before and after the First World War. All were agreed that agricultural higher education in the United States was more practically oriented than in Germany.[9] (This is not to say that American institutions were uniformly so. As in Britain, agriculture was taught in multi-faculty universities as well as in specialised colleges, and the latter once again were more practical in character: the curriculum was more vocationally oriented, extension played a larger role, and research had a stronger economic emphasis than in the universities.)[10]

The overall practical orientation of American institutions was sustained by strong organisational links to the farming community. German visitors frequently remarked that, unlike in Germany, the American colleges were formally attached to both experiment stations and to extension services. Indeed academic staff were expected to devote part of their work to extension which meant that they were well informed about the problems faced by local farmers (partly because the latter were not shy about voicing their needs).[11] The importance of these links was also noticed by the Americans, C.B. Hutchison and Albert Mann, during their tour of European agricultural colleges in 1927. Since in Europe neither the kinds of routine testing conducted at experiment stations (eg, of plant varieties or fertilisers) nor extension work was a responsibility of the colleges, they noted,

9 Hutchison and Mann, 'Higher Education in Agriculture'; von Ruemker and Tschermak, *Landwirtschaftliche Studien*, 128–129, 132; Tschermak, *Leben und Wirken*, 152–153; Aereboe, *Wirtschaft und Kultur*; Roemer, *Beobachtungen auf dem Gebiete des Ackerbaues*, 117–118.

10 Shepardson, *Agricultural Education*, 98–99.

11 Eg, Sohn, 'Das landwirtschaftliche Beratungswesen'; Aereboe, *Jugendlust*, 196–198; Roemer, *Beobachtungen auf dem Gebiete*, 109–110. On Max Maercker's visit in 1893 he was struck by the close contact between American experiment stations and farmers (Finlay, *Science, Practice and Politics*, 374–375).

> The university professor is a freer agent, at liberty to direct his research in any direction or at any level he may choose [...]. The [...] professor is thus generally without pressure from constituents and feels less urge to deliver immediately consumable results [...]. His research laboratory, therefore, allows the quiet pursuit of knowledge in whatever direction his inclinations and [...] abilities lead him, and enables him to devote himself to the broader problems. He is, however, much less a factor in current agricultural progress than the American agricultural college teacher.[12]

Given the orientation of American institutions, it is perhaps not surprising that they enjoyed a substantial degree of respect from farmers, a fact which much impressed the Germans.[13]

The practical orientation of American education may also, however, have been shaped by the nature of agricultural practice. For in contrast with western Europe, as German visitors noted, American agriculture in the late 19th and early 20th centuries was, for the most part, extensive rather than intensive.[14] Land was cheap and abundant, and profits could be reaped with only modest capital investment. Thus farmers were more concerned to lower production costs (eg, via mechanisation) than to increase yields, and in consequence before the First World War, at least, 'scientific agriculture' had relatively limited appeal.[15]

Not all interested parties, however, were sanguine about this orientation. One group of organisations which actively sought to promote a stronger science-orientation in American agricultural education

12 Hutchison and Mann, 'Higher Education in Agriculture', 13.

13 Roemer, 'Variabilitaetsstudien' (1910), 397–398; Aereboe, *Wirtschaft und Kultur*, 34; Wohltmann, *Landwirtschaftliche Reisestudien*, 90–105; E. Mangold reported on his trip to the U.S. in his letter to rector of Berlin Agric. College, 1 Oct 1929, GStaA, R87B, Nr. 12548.

14 Wohltmann, *Landwirtschaftliche Reisestudien*, 106–122; Roemer, 'Referat von Professor Th. Roemer – Halle ueber die Amerikareise', 27 Oct 1925, UA-Halle, Rep 34G, Nr. 29.

15 Danbom has argued that the champions of scientific agriculture before the First World War were urban groups primarily concerned with promoting an industrial economy (allied with agricultural scientists) rather than the farming community, most of whom were not persuaded that increasing productivity through hi-tech methods was in their collective interest (Danbom, *Resisted Revolution*).

were the Rockefeller philanthropies. In the mid-1920s, for example, the International Education Board funded Albert Mann (dean of Cornell's School of Agriculture and head of the IEB's section for agricultural education) and C.B. Hutchison (subsequently dean of agriculture at the University of California) to make a study of European agricultural education. Hutchison's 'prime purpose in life', as he told a historian later, was to 'strengthen the scientific basis for agriculture'.[16] In their report to the IEB, Hutchison and Mann declared that governments had so far been too concerned with immediate utility and insufficiently interested in promoting basic sciences within agriculture 'which become finally the preconditions of further advance.' There was thus 'a very real need and opportunity for private benevolence to [...] foster [...] the higher and more remote, but now already imperative, fundamental and theoretical aspects of agricultural science.' [17] This perspective was evident in Hutchison's comments on various German institutions. Visiting Hohenheim and Weihenstephan he felt it was 'undoubtedly a fundamental weakness of these two institutions' that basic sciences and economics there were taught with an agricultural slant.[18] And when an applied botanist elsewhere told him that the most important contributions to plant production had been made by botanists, Hutchison commented 'probably true'.[19] Nor was the IEB the only Rockefeller philanthropy to view agricultural science in this light. A report on American agricultural education prepared for the General Education Board a few years later took much the same line.[20]

16 C.B. Hutchison, 'Oral History', 91, C.B. Hutchison papers.
17 Hutchison and Mann, 'Higher Education in Agriculture', p. 9.
18 p. 260, box 1, file 33, C.B. Hutchison papers.
19 p. 121, box 1, file 33, C.B. Hutchison papers.
20 Shepardson, *Agricultural Education.* There had been too much emphasis, Shepardson wrote, upon extension and practical farming skills and not enough upon scholarship. Research staff needed more time free of calls for 'service', and 'pure' science required more support (77, 79). Underpinning his recommendations was a classical vision of the agricultural sciences in which 'technology is applied science'. Physics, chemistry and biology were said to provide the foundation of the agricultural sciences while nutrition, pathology, and soil sciences were 'derivative subjects'(69). Yet further away from the basic sciences were agronomy and animal husbandry, 'fields of practical application'

Moreover the Rockefeller philanthropies put their money where their mouth was. For example, when the IEB awarded a large development grant to Cambridge University's School of Agriculture in 1928, practical fields like animal pathology and nutrition were excluded; the bulk of the funding was earmarked for plant physiology and genetics, and the Board wanted to encourage closer ties between the School and the University's science departments.[21] But perhaps the largest and best-known of the philanthropies' ventures in agriculture was the Rockefeller Foundation's funding of the 'Green Revolution' from the 1940s.[22] The conception of agriculture which informed their programmes was reminiscent of Julius Kuhn's: 'Agriculture is nothing more than the application of the principles of biology and other natural sciences to the art of growing food.'[23] In accord with this view, the Foundation's strategy for solving 'the world hunger problem' was to promote in Third World countries the kind of science-based plant-breeding which in the United States had produced high-yielding cereals-varieties. The strategy was attractive to the Foundation partly because of its long-standing interest in science-based agriculture but also because the Green Revolution seemed to be a purely technical solution to an economic problem: it promised to increase the production of food without alterring its distribution or in other ways interfering with the social relations of agriculture. As we have seen in the German case, however, such science-based intensive cultivation methods were more readily taken up by large landowners than by peasants. Indeed, the reason why plant-breeding stations were established at Hohenheim and Weihenstephan around the turn of the century was to counteract this tendency so as to render plant-breeding technology more accessible to the small farmer (Chapter Four). The outcome of the Green Revolution was thus predictable. New high-

(69–70) which were empirical in character and whose emphasis was on results. Increasing agricultural production, the author claimed, 'has never been done nor can it be done except through the discoveries of science'(72). 'The understanding of nature has always led to its control'(79).

21 Kohler, *Partners in Science*, 187; see also Ede, 'School of Agriculture'.

22 Jennings, *Foundations*; Anderson et al., *Rice Science*; Perkins, *Geopolitics*.

23 An advisory committee to the Rockefeller Foundation in 1951, cited in Anderson, 'Origins of the International Rice Research Institute', 64.

yielding rice and wheat varieties were developed (with their characteristic requirements for fertiliser, irrigation and mechanised harvesting) which were successfully adopted by large landowners, and the increased yields permitted several countries to develop a cereals-surplus. The new varieties, however, were not adopted by most peasants who could not afford either the seed or the growing conditions which it requires. Thus the story of the Green Revolution illustrates the more general point that science-oriented approaches to agricultural problems have had a decidedly selective impact upon the agricultural economy, benefitting industry and large landowners rather more than most farmers.

If American agricultural education started out with a markedly practical bent, however, it has not remained unchanged. Although established to serve the farming community, the agricultural universities have grown steadily more science-oriented during the 20th century. The trend began as early as 1906 when the Adams Act allocated funds to the agricultural experiment stations which were earmarked for 'original research' rather than for the practical testing of fertiliser, seeds and soil samples which had been of more immediate interest to farmers. A few years later the Smith-Lever Act (1914) established a network of county agents to handle matters of extension, thus freeing academics of some of that responsibility.[24] During the 1920s, to be sure, the drift away from practice was momentarily halted. When postwar overproduction lowered prices, only to be lowered even further by depression during the 1930s, farmers demanded that agricultural research be redirected, and Congress allocated more funding for economic and sociological study of agricultural problems.[25] But with the outbreak of the Second World War, increases in productivity again became important, and the trend resumed. According to some authors, whereas in the 1950s the agricultural universities' links with the farming community were still relatively strong – eg, through extension advice and the release of newly-developed commodities to private firms and growers' associations – by the 1980s it was mainly large farmers who obtained

24 Rosenberg, *No Other Gods*; Marcus 'Wisdom of the body politic'.
25 Danbom, 'Publicly sponsored agricultural research', 118–120.

advice. Universities had largely stopped releasing commodities, and they were beginning to conduct more basic research in the life sciences.[26]

Academic drift has nowhere been more evident than in plant-breeding where since the 1920s, under pressure from the private sector, the universities' release of new finished varieties onto the market has gradually declined.[27] Initially this meant stopping the distribution of free seed to farmers; subsequently academic breeders were persuaded to concentrate on developing parent materials which were turned over to private seed companies for finishing and sale. In the 1950s the pressure upon the universities intensified, especially from *large* seed companies who, unlike their smaller counterparts, could afford to conduct their own in-house breeding research. Since the 1970s a series of government reports have taken agricultural universities to task for the neglect of basic research in plant-genetics, arguing that such research provides a crucial foundation for the plant biotechnology industry and that the development of new varieties should be left to the latter sector which has both the necessary research capacity and the marketing know-how. The private sector, of course, chimed in, and the consequence, evident already during the 1980s, was a shift at the universities away from applied plant-breeding toward plant molecular genetics.[28] Unless this trend is countered by sustained lobbying by farming, rural, environmental or other organ-

26 Buttel et al., 'Industry/land grant university relationships'; cf. Busch & Lacy, *Science, Agriculture and the Politics*, chp. 11 and passim.

27 Busch et al., *Plants, Power and Profit*, 192; Kloppenburg, *First the Seed*, 71, 108–110.

28 Busch et al., *Plants, Power and Profit*, 91–94; Kloppenburg, *First the Seed*, 235–240; Kenney, *Biotechnology*, 230–236. Moreover, the impact of this disciplinary shift upon academisation in the United States is likely to be augmented by the fact that at American agricultural universities substantially more plant-breeders than those in molecular genetics come from farm families (Busch et al., *Plants, Power and Profit*, 90). In Britain the Thatcher government pushed through a similar shift – opposed by farmers' organisations – by radically cutting back spending on some public-sector institutes and selling others to multinational companies (Anon., 'The worst is still to come'; Hadlington, 'Unilever purchases').

isations, the day is rapidly approaching when the breeding of new varieties will disappear from the universities altogether.

From this literature, therefore, it appears that the causes of academic drift in American agricultural research are to be found in institutional changes introduced by legislation, pressure from particular sectors of the agricultural economy and other interest groups, and a changing pattern of government funding (in which that from state governments attuned to farmers' concerns has given way to that from federal agencies which are little interested in agriculture, such as the National Science Foundation and National Institutes of Health). Although to my knowledge no one has yet attempted a systematic explanation for this trend, it may well be amenable to the kind of analysis I have developed here.

Rethinking the history of engineering, medical and management education

As I noted at the outset, the processes addressed in this book are unlikely to be peculiar to agricultural higher education. Since all areas of the practical sciences are institutionally situated between basic disciplines and the economy (broadly defined), we would expect to find 'science-oriented' and 'practice-oriented' forms of education there, too, as well as a tendency to academic drift. If that is so, then the approach developed here should be applicable to the entire range of the practical sciences. Let us consider what is known about the history of education in engineering, medicine and management sciences.

Many historians of 19th and early 20th century engineering – in Germany, the United States, France and Britain – have recognised the existence of broadly distinct forms of engineering knowledge, practice and education which correspond to those I have termed the 'scientific' and the 'practical'. Moreover, the concept of 'academisation' was first developed by historians of engineering and has been demonstrated in all of these countries. The question which must be asked, however, is whether historians of engineering have succeeded in explaining the emergence of these orientations.

Twenty years ago, in a review of the literature on higher engineering education in Germany, Wolfgang Koenig called for more comparative studies which were informed by specific research questions.[29] It seems to me that, by and large, that need still exists, not just for institutions in Germany but elsewhere as well. Studies of the development of engineering knowledge have for the most part taken one of two forms. On the one hand there are excellent case studies of 'science-' or 'practice-oriented' curricula (less often of research) at particular institutions, among them John Servos on MIT, John Weiss on the Ecole Centrale, and David Channell's study of Rankine.[30] Occasionally such studies have been comparative, but these have either focussed upon general institutional history while saying little about the details of teaching and research,[31] or they have been disciplinary histories which make little attempt to situate curricula in institutional contexts.[32] Nor have the authors of even the few cross-national studies used comparison to venture any generalisations about the constitution of academic technical knowledge.[33] On the other hand, a number of historians have addressed important general trends such as academisation, some of them providing descriptive overviews[34] and others offerring contextualised explanations.[35] But none

29 Koenig, 'Stand und Aufgaben'.

30 Servos, 'Industrial relations of science'; Weiss, *Making of Technological Man*; Channell, 'The harmony of theory'.

31 eg, Sinclair, 'Harvard, MIT'; Shinn, 'Reactionary technologists'.

32 eg, Dienel, 'Der Muenchener Weg'; Koenig, *Kuenstler und Strichezieher*.

33 Eg, Lundgreen, 'Engineering education'; Locke, *End of Practical Man*. Eda Kranakis's study of bridge-building engineers in 19th century France and the United States is an exemplary analysis of the design-process within economic, social and organisational contexts, but it is not primarily concerned with higher education (Kranakis, *Constructing a Bridge*).

34 eg, Braun, 'Methodenprobleme'; Kline, 'Construing "technology"'.

35 eg, Manegold, 'Technology academised'; Zweckbronner, 'Die Ingenieurwissenschaften'; Seely, 'Research, engineering and science'; Servos, 'Industrial relations of science'; Day, *Education for the Industrial World*; Shinn, 'From "corps" to "profession"'.

have extracted general concepts or hypotheses which might be applicable elsewhere, whether in other countries or in other historical periods. Thus we still lack models for explaining why 'science-' and 'practice-oriented' forms of engineering education developed when and where they did or why the process of academisation has so often occurred.[36]

The literature on medical knowledge and education is rather similar. To place medicine alongside agriculture and engineering within the domain of the practical sciences is perhaps slightly unusual in that history of technology and history of medicine are conventionally regarded as distinct fields. In addition, it would be hard to argue that scholars in medical faculties have suffered a status problem (over the last century, at least), within either academe or the wider society, comparable to that experienced by their colleagues in agriculture or engineering. Nevertheless, it doesn't take much effort to see that academic medical knowledge has also been situated between 'sciences' like anatomy and physiology and the actual 'practice' of medicine. Indeed, in the 19th and early 20th centuries academic engineers who sought to get their subject accepted within the university – like their counterparts in agriculture – sometimes stressed its parallels with medicine.[37] In late 19th century Britain and the United States this social and cognitive ambivalence was particularly obvious. Originally a craft-based body of knowledge derived from clinical experience and reproduced in part via a form of apprenticeship, medicine was targeted for reform at this time by advocates of a 'scientific' medicine, an approach developed somewhat earlier in Germany where laboratory instruction was central. The reformers proposed to place medical knowledge on new foundations so that doctors would have a fuller understanding of disease, and medical knowledge would be continually renewed via laboratory-based research. Though initially resisted within the profession, the newer scientific approach eventually found widespread support in Britain and the U.S. as can be seen in the structure of the medical curriculum nearly everywhere for

36 I have discussed this literature at length in 'Engineering education between science and practice: rethinking the historiography' (forthcoming).

37 Marsden, 'Engineering science in Glasgow'; Morrell, *Science at Oxford*, 100.

most of the twentieth century: two years of pre-clinical sciences provided the 'foundation' before the student encountered the hospital ward.

This late 19th century 'laboratory revolution' in medicine poses many of the same historiographical questions already discussed for agriculture and engineering: Why were the 'clinical' and the 'scientific' conceptions of medical knowledge institutionalised at some places but not others? And why has the orientation of medical education shifted over the last century from 'clinical' to 'scientific' (and more recently back again)? Several historians have looked at the confrontation between the older clinical and the newer scientific conceptions, whether in a specific hospital, in the clash between the Medical Research Council and other medical organisations, or in debates over the foundations of therapeutic knowledge.[38] Not all of them have been interested in explaining why physicians have endorsed such different visions, but some have, and they are agreed that adequate explanations do not yet exist.[39]

Answering this question would seem to require close contextual studies of groups of medics who shared one or other vision of medical knowledge and of particular institutions in which one vision predominated. Two nice examples of this kind are Chris Lawrence's study of British advocates of 'hospital medicine' in the decades around 1900 and Steve Sturdy's contrasting study of those Sheffield clinicians who championed laboratory medicine about the same time.[40] The existence of such case studies naturally poses the more general question as to whether any pattern can be found in the kinds of institution in which one or other conception of medicine predominated. Since case studies of this kind are still rare, there have been only

38 eg, Jacyna, 'The lab and the clinic'; Austoker, 'Walter Morley Fletcher'; Warner, *Therapeutic Perspective*, chps. 8 and 9; Warner, 'Ideals of science'.

39 On the insufficiency of age and training see Warner, *Therapeutic Perspective*, 274; Warner, 'The history of science', 179, 182.

40 Lawrence, 'Incommunicable knowledge'; Sturdy, 'Political economy'. Another critique of the idea that medical practice should be derived from science was developed at about the same time by philosophically inclined physicians in Poland, of whom the best known to historians of medicine and science is Ludwik Fleck (Loewy, *Medical Acts and Medical Facts*.).

tentative suggestions in the literature thus far, among them that the laboratory tradition in Britain may have been stronger ca. 1914 in provincial medical schools than in London's hospital-based schools.[41] Should this hypothesis be borne out by future work, such a pattern would demand explanation. And for that purpose comparative analysis – of a kind which at the moment seems not to exist – will be necessary.

The other literature of interest is that which deals with the history of medical education. The key questions which need to be asked here, too, are: what processes have shaped the medical curriculum over time, and why have curricula at some medical schools been more science-based than at others? As far as the former is concerned, it might seem obvious to anyone with experience of academe that curricula are generally the outcome of struggle within academic institutions (although external groups may also be important players in this game). Nevertheless, several well-known accounts of the evolution of medical education (in the United States) have practically nothing to say about the politics of the curriculum. The authors seem little interested in the alternative curricula developed at some medical schools and take it for granted that the science-based curriculum of the 20th century represents the culmination of a natural and largely uncontested process.[42] As we have seen, however, there was nothing at

41 Lawrence, *Medicine in the Making*, 72; Sturdy, 'Political economy', 158. By the interwar period the situation seems to have been much less clearcut. On the one hand there were new MRC-funded units for clinical research at several London hospital-schools, but on the other, a series of attempts to establish full-time research posts at Manchester faltered when the occupants either abandoned research or left after a few years. Only from the late 1930s did clinical research begin to make inroads there (Valier, 'Politics of Scientific Medicine').

42 eg, Kaufman, 'American medical education'; Ludmerer, *Learning to Heal*, cf. 4; 266; Bonner, *Becoming a Physician*, esp. 268–278, 289; 347. Both Kaufman and Ludmerer briefly acknowledge curricular 'experiments' at one or more medical schools which were designed to overcome the gulf between the clinical and pre-clinical years but do not elaborate (Kaufman, 'American medical education', 26–27; Ludmerer, *Learning to Heal*, 262). It is a great strength of William Rothstein's book, on the other hand, that it demonstrates so clearly how medical educators were divided over questions of 'science' versus 'practice' throughout the 20th century (Rothstein, *American Medical Schools*, esp.

all 'natural' about the process of academisation in agricultural education; in certain contexts, for example, economic constitutiencies objected vigourously so as to block the process. Any satisfactory account of the academisation of medical education, therefore, will need to look, not only at the aims of medical academics, but also at forces external to the medical schools which may have facilitated that process, such as an abundance of government funding for basic medical sciences or foundation-backed reform movements.[43]

Finally, consider the management sciences. The literature on business education is still relatively small, but it suggests a number of parallels with other areas of the practical sciences. In Germany at least, business education was caught in the crossfire between the worlds of science and practice.[44] Neither the universities on the one hand, nor most of the business community on the other, were convinced of the need for the commercial colleges. But once they were established, the pressures to conform to academic norms were much in evidence, not least in the curriculum. Before the First World War when no coherent discipline existed which staff at business schools could call their own, a certain amount of curricular diversity was available. At Leipzig and Cologne, as Keith Tribe has shown, commercial subjects (*Handelswissenschaften*) were subsidiary while economics was regarded as the core of the curriculum, partly because it was well-recognised within the academic world. In Berlin, by contrast, the core consisted of commercial subjects with economics and law as

chps. 7, 8, 11 and 16) though the reasons behind that diversity are left unexamined.

43 Rothstein, *American Medical Schools*. Susan Lawrence's overview of the history of medical education displays more sensitivity to the contested character of the curriculum, though none of the post-1914 challenges to the curriculum which she discusses concern the proper balance between scientific and clinical elements in training (Lawrence, 'Medical education').

44 Early 20th century German management scientists (Betriebswirte) were aware of these conflicting pressures – seeking recognition from the educated middle-class as a respectable academic discipline while trying to be useful to the commercial middle-classes – and described themselves as 'fight[ing] a war on two fronts' (Franz, 'Betriebswirte', 259–262).

subsidiary disciplines.[45] On the eve of the war attempts to develop a distinctive discipline which could constitute the core of commercial education were underway, but its practitioners were concerned that their young discipline would never be accepted within academe if they took anything so coarse as profit as the measure of a firm's success. So they declared the measure instead to be 'efficiency' or 'productivity' (*Wirtschaftlichkeit*). And by the early 1920s the new 'business economics' was well enough regarded by university economists that they began to offer it to their own students. But status-insecurity continued to affect the discipline. Heike Franz has argued, for example, that one particular theoretical development was resisted by business economists (even after 1945) because it would have made their discipline directly useful to businessmen, thereby [sic] undermining its credentials as impartial scholarship.[46] As one might expect, given these examples, the pressures upon business educationists to academise their subject were substantial (though the process was opposed by some members of the business community), culminating in 1930 with the right to award the doctorate.[47] Thus higher business education displays a number of close resemblances to the other prac-

45 Tribe, *Strategies of Economic Order*, chp. 5. By 1914 the new business economics had become central to the curricula at all of these institutions, but quite why their initial starting points were so distinct is not clear. Tribe hints that different organisational features may have been responsible, making for tighter links between the college and the business community in Berlin than in Cologne. But the issue is not explored at any length since his main interest is the evolution of business economics rather than the institutional history of commercial education as such. Redlich ('Academic education for business') also notes the curricular contrast between Berlin and the others but attributes it simply to the insight of Berlin's founding father, Ignaz Jastrow. Franz notes that by 1916 the emphasis upon business economics was particularly strong at Berlin but offers no explanation (Franz, *Zwischen Markt und Profession*, 66, cf 54).

46 Franz, *Zwischen Markt und Profession*, 100–103.

47 Franz, *Zwischen Markt und Profession*, esp. chp. 4 on academisation; see also Kirchgaessner, 'Die Gruendung der Handelshochschulen'; Locke, *End of Practical Man*, 159–160, 202–203; Lindenfeld, 'Professionalisation of applied economics'; Tribe, 'Business education'; Franz, 'Betriebswirte'.

tical sciences, suggesting that it, too, could be illuminated by the model.

The foregoing discussion of these literatures has been brief, but I hope enough has been said to indicate that the model developed in this book ought to be applicable across the whole range of the practical sciences. Whether or not that proves to be the case, there is a lot to be said for historians of technology, medicine and management occasionally getting together to discuss common problems.

Implications for the reform of technical and professional education

Since 1945 there has been repeated criticism in several areas of the practical sciences that higher education was not producing the kinds of graduates which are necessary. In engineering, for example, French industrialists complained during the 1990s about the lack of practically-oriented engineers while in Britain the Finniston Report (and others since) have voiced similar discontent. In the United States engineering colleges have been concerned in recent years about the continuing gap between what academe supplies and industry demands. Closing the gap, however, is not easy, partly because some academic engineers continue to insist on calculus and higher mathematics in the curriculum while admitting that these may be of little relevance for job performance.[48]

In medicine criticisms of an excessively science-oriented curriculum in the United States date from at least the 1930s. In the 1980s both the Association of American Medical Colleges and the American Medical Association noted with concern that staff were not devoting much attention to the provision of general medical education. Accord-

48 Grelon, 'Die deutschen Ingenieure', 383; Committee of Inquiry, *Engineering our Future*; Pollard, *Education and Training*; Williams, 'All that is solid'; Seely, '"Imbalance" of theory', 57. On mathematics in the curriculum see Hacker, 'Mathematization of engineering'.

ing to Rothstein, the main impetus for improving training in primary care over the last thirty years has come, not from the medical schools, but from state and federal governments. Nor, it would appear, have medical schools responded to such initiatives with much enthusiasm.[49] In Britain, similarly, there were sporadic attempts at curricular reform between the wars though with little impact. Only since the 1990s have major changes been implemented at some medical schools, usually where pre-clinical departments' lack of interest in educational matters – and power over the curriculum – could be overridden.[50]

And as we saw above, academic drift has been underway for some time in American agricultural education. Since the 1960s a vigorous debate has taken place in the United States – intensified by the rash of farm closures during the 1980s – about the consequences of this trend.[51] A number of critics have argued that in promoting 'scientific agriculture' in the interests of increased productivity, the agricultural faculties of the state universities abandoned their historic mission to serve the farming community as a whole and now aid only a minority of farmers along with agricultural firms. According to the critics, researchers in the universities are no longer as responsive to their extension departments; promotion is increasingly awarded on the basis of conventional publications rather than bulletins aimed at the farm community; and funding is no longer as geared to agricultural problems as it once was. Meanwhile the intensive cultivation championed by the universities has devastated rural communities as the majority of (less productive) farms has been displaced by a small number of large ones using ever more advanced hi-tech methods, which in turn have cost farm labourers their jobs. Nor does this form of agriculture further the interests of the general public which pays for these universities. Pesticides threaten the health of consumers and farm workers; fertiliser-runoff and soil erosion spoils the environment; agri-

49 Rothstein, *American Medical Schools*, 300–302, 312–313, 326–327.

50 Mahadevan, *Creating Tomorrow's Doctors*.

51 Hightower, *Hard Tomatoes*; Hadwiger, *Politics of Agricultural Research*; Busch and Lacy, *Science, Agriculture & Politics*; see also the papers in part 7 of Busch and Lacy, *Agricultural Scientific Enterprise*; Buttel and Busch, 'Public agricultural research system'.

cultural research seems more concerned with increasing productive efficiency than nutritive value; and critics question the point of increasing efficiency yet further when there are already huge food-surpluses.[52] These charges obviously raise fundamental questions about the proper responsibilities and future role of agricultural education and research.

In all three of these areas it is noticeable that the pressure for reform is coming from without; educational institutions themselves are evidently part of the problem rather than part of the solution.[53] But clearly reform is possible; the fact that not all engineering colleges nor medical schools, for example, have gone down the same science-oriented path indicates that academic drift is certainly not inevitable.[54] The question, therefore, is what kinds of organisational changes would be necessary in order to keep technical and professional education 'on track' in future? More particularly, what might be the implications of this book for educational policy?

It is a truism that history never repeats itself; the future never simply reproduces the past. Thus it would be naïve to expect history to provide a blueprint for action. What it can do instead is to provide tools for thinking, a way of looking at the problem which may suggest possible solutions. Reflecting on the model developed in the foregoing chapters, it seems to me that there are three general points which might be borne in mind when establishing new educational institutions or reorganising old ones.

52 Since the 1980s a similar debate has taken place in the Netherlands, as a result of which agricultural institutions have been urged to give more attention to much the same range of public concerns (Maat, *Science Cultivating Practice*, 219ff).

53 On American agricultural colleges' failure to address the farm crisis or to develop proposals for their own reform, see Hadwiger, *Politics of Agricultural Research*; Buttel and Busch, 'Public agricultural research system', 318.

54 That academisation during the postwar period has not been universal is evident from the claim that engineering curricula in Germany, Japan and the Netherlands, in contrast with those in the United States, have not become so remote from industrial requirements (Ferguson, *Engineering and Mind's Eye*, 159–168).

The first should be fairly obvious, though it is ignored surprisingly often: that geography matters. It is already recognised, for example, that medical schools are best situated in larger urban areas so that students can become acquainted with (and staff can investigate) the widest range of health problems.[55] Engineering colleges, similarly, need to be located near industrial areas so that linkages such as consulting and sponsorship are easier to arrange. From this point of view smaller cathedral towns – the site of so many new British universities ca. 1970 – while offering a pleasant environment for work and study, are scarcely appropriate sites for technical education. As far as higher agricultural education is concerned, it is unlikely that new colleges will be established in the near future. But for those existing ones which are beginning to reorient their research and teaching toward a larger public, geographical location ought to be a factor in deciding whether to aim for rural communities or urban consumers.

Second, as we have seen, the development of educational institutions has been shaped by pressures from both government and user-groups in the wider economy. After the First World War the Prussian and Wuerttemberg Ministries of Education tried to increase agricultural bodies' influence by establishing advisory boards at the agricultural colleges. Although these were understandably not popular with academic staff, it is hard to avoid the conclusion that something similar is necessary today in order to keep such institutions responsive to their constituencies. Defining the membership of such boards, of course, is a political issue and would need to ensure accountability to relevant publics at the local, regional and national levels. The representatives of 'the public interest' on an engineering college's board, for example, should not be restricted to industrialists, and it would be more important to have local general practitioners and laypeople on a medical college's board than to include consultants or non-medical academics.[56] Where higher education is in the hands of the state,

55 This issue is especially topical in Britain at the moment when a number of new medical schools are being established in order to meet the increased demand for physicians.

56 This issue arises, for example, through recent recommendations for the reform of British veterinary education (Royal College of Veterinary Surgeons, 'Report

boards might advise the relevant ministries, but ultimate authority naturally resides with the ministers themselves, and it is abundantly clear from the case studies discussed in earlier chapters that ministerial action was able to block academisation, just as indifference could inadvertently further it. One way in which the state affects academisation, of course, is through targeted funding. But if academisation is to be discouraged, the rules governing such funding must be sensible. Rewarding colleges primarily for publishing in conventional academic journals – as has been the practice in British 'research assessment exercises' over the last two decades – sends the wrong message. Other forms of research or 'outreach' activity such as consulting, patenting, publishing in trade journals, or collaboration with user-groups must also receive recognition.

Finally, one ought to consider how the structure of the higher education system itself might be altered. At the heart of academisation, according to the model I have proposed, is the perceived higher status of the basic sciences. As pernicious as this value is, however, it seems quite unrealistic to hope that it can be changed in our lifetime. For in the western tradition the subordination of hand to head and of 'practice' to 'theory' is much too deep-rooted to be shifted easily. The British engineering professor, W.J.M. Rankine, railed against it in the mid-19th century, and in the 18th Diderot 'seldom [lost] an opportunity to strike a contrast between the pathetic and appealing humility of technique and the haughty arrogance of mathematical abstraction'. In classical antiquity, too, those branches of knowledge which aimed at utility rather than knowledge for its own sake, were generally deemed inferior, and a similar hierarchy of

of the Committee of Inquiry'). The Report complains of the overly practical curriculum and the lack of research opportunities at most veterinary colleges (and the same is said to apply to dental education). Before introducing any reforms – which would effectively bring about academisation of the colleges – however, any government must ascertain whether the RCVS recommendations in fact meet the needs of most of those groups with an interest in veterinary education.

knowledges was common to ancient Indian and, for the most part, Chinese traditions.[57]

But if these values are so long-standing and widespread, educational reform will require institutional changes which can counteract them.[58] One way to proceed would be to reduce the opportunities for this prejudice to be expressed. Recall that at both Halle and Breslau agricultural staff (who were in a small minority there) ran up against a good deal of condescension and scepticism from their colleagues in the natural sciences and humanities, and there are signs that the situation at the engineering college in Munich may have been similar. Were one to try to reduce this damaging social interaction by simply removing the agricultural sciences from such large multi-faculty institutions, however, the consequences could be serious. For as I argued in Chapter Two, the development of a strong science-orientation has not *invariably* been a bad thing. In some areas of engineering, as in plant-breeding at Halle under Roemer, such an orientation was very productive of innovative and useful work. So how is an institution to maintain the possibility of fruitful connections between the practical sciences and basic sciences while keeping the destructive snobbery of head-and-hand at bay? An obvious solution would be to place each of the technical sciences in specialised colleges which contained posts or departments dedicated to the relevant basic subjects. This is in fact just how the former German engineering colleges were organised, and it is the direction in which German agricultural colleges were moving by the First World War. In such technical institutions basic scientists

57 On Rankine see Calvert, *Mechanical Engineer in America*, 42. Diderot is quoted in Gillispie, 'The Encyclopedie', 270. On antiquity see Stokes, *Pasteur's Quadrant*, 27–30; on India and China see Bray, *Technology and Gender*, 41–44, and Sivin, 'Why the Scientific Revolution'.

58 It is interesting in this regard to note Hightower's recommendations for the reform of American agricultural research (*Hard Tomatoes*, chp. 9). Many of them are pitched at the level of national policy-making (eg, public inquiries, congressional committees) while relatively few concern the structure of agricultural institutions or the higher education system more generally. Nothing is said, for example, about changing the reward systems or research-funding arrangements for agricultural scientists, nor about restructuring the extension service so as to give it more influence upon research agendas.

constituted a minority 'tail' which was in no position to wag the dog. Moreover a condition of these scientists' appointment was that they were prepared to explore possible links between their own disciplines and related technical ones. In this way colleges of engineering, medicine, management or agriculture could potentially reap the cognitive benefits of the sciences' proximity while not suffering the institutional disadvantages.[59]

Creating specialised technical institutions, however, would not be sufficient on its own to forestall academisation. For as we have seen, staff at the agricultural colleges, like those at engineering colleges in the late 19th century, still fought hard to raise the formal academic status of their institutions and in so doing were prepared to further the process by shifting their activities closer to 'science' (a clear example of which was the Agricultural College at Berlin). This suggests that education ministries will also be obliged to introduce measures which undermine the formation of academic status hierarchies by assuring formal equality among the different sectors of higher education. As we have seen, this is what German education ministries largely failed to do. Even after they had granted German engineering colleges the much-coveted right to award the doctorate at the turn of the century, a variety of less obvious organisational differences and privileges persisted between the colleges and the universities which continued to provoke resentment among academic engineers into the 1920s (and the situation was little different in the agricultural colleges).

Even today there are signs that education ministries have real difficulty in conceiving of higher education systems in non-hierarchical ways. In Germany, for example, while most of the old engineering colleges were upgraded over the last few decades to 'technical

59 Studies of academic drift in the medical curriculum indicate that one major reason for this process has been that staff in the pre-clinical sciences are responsible for teaching, not only medical students, but also general biological sciences (Rothstein, *American Medical Schools*; Mahadevan, *Creating Tomorrow's Doctors*). Moreover, their research is funded largely by bodies responsible, not for medical research, but for basic biology. Under the circumstances, who can be surprised that these staff have been more interested in their own research and graduate students than in teaching medical students?

universities', there is still a lower rung on the academic status-hierarchy occupied by 'technical colleges' (*Fachhochschulen*). These colleges have been much praised by both industry and the national advisory body on higher education (*Wissenschaftsrat*) because their graduates are more practically-oriented than university graduates, and there have been frequent calls for more of them to be established. Unfortunately, however, students have not been enrolling at these colleges in increasing numbers as had been hoped, in part because graduates of the colleges – unlike university graduates – are not permitted to apply for jobs in the higher civil service. Moreover there have been difficulties in filling academic positions at the colleges with highly qualified staff, no doubt because staff there are not as well paid as those at universities, have much heavier teaching loads, and have relatively limited research facilities.[60] The state has thus been sending a clear message: that the technical colleges, despite their admirable qualities, are second-class institutions. Judging from the history of technical education, the likely outcome is that staff in these colleges will press for the kinds of institutional changes which they believe will qualify them for university-status – academisation revisited – and will, in the process, erode the very qualities which currently make them so attractive.[61] If governments want to preserve this kind of institutional diversity, therefore, they must dismantle the stratification of higher education so that institutions can be different but equal. Allowing formal differences of status to persist is asking for trouble.

These are some of the ways in which history might inform policy. Obviously the model I have developed here is hardly the last word on the dynamics of institutional change. Theories get better (or are eventually abandoned, as the case may be) the more they are used. And applying them to a variety of different circumstances is particularly informative since it reveals their weakspots and limits. That

60 Wissenschaftsrat, 'Empfehlungen', 93–95 and 147–150; cf. Spiewak, 'Klassenkampf'.

61 This seems to be what happened at comparable British institutions (the polytechnic colleges) between their creation in the 1960s and their upgrading to universities in 1992.

is why I hope that historians and educationists concerned with all of the practical sciences will find this book of interest.

* * * *

Bibliography

Archival Sources

BayHStA – Bayerisches Hauptstaatsarchiv, Munich.
GStaA – Geheimes Staatsarchiv Preussischer Kulturbesitz, Berlin.
RAC – Rockefeller Archive Center, Tarrytown, New York.
R21 (Reichsministerium fuer Wissenschaft, Erziehung und Volksbildung), II. Anhang (Erfassung der Hochschullehrer, 1934–44), Bundesarchiv Koblenz.
TUM-Archiv – main Archive of the Technische Universitaet Muenchen.
UA-Giessen – Archive of the Universitaet Giessen.
UA-Halle – Archive of the Universitaet Halle.
UA-Hohen – Archive of the Universitaet Hohenheim
UA-Weih – Archive of the Technische Universitaet Muenchen, Verwaltungsstelle Weihenstephan, Freising.
Wurtt HStaA – Wuerttembergisches Hauptstaatsarchiv, Stuttgart.
Friedrich Althoff papers, Geheimes Staatsarchiv, Berlin.
William Bateson papers, John Innes Centre, Norwich, UK.
Emil Fischer papers, Bancroft Library, University of California, Berkeley.
C.B. Hutchison papers, Department of Special Collections, Shields Library, University of California, Davis.
Hermann Nilsson-Ehle papers, University Library, Lund, Sweden.
Elisabeth Schiemann papers, Staatsbibliothek Preussischer Kulturbesitz, Berlin.

Printed Primary Sources

Bericht ueber die Bayerische Akademie fuer Landwirtschaft und Brauerei in Weihenstephan, 1899/00 to 1910/11.

Bericht ueber die Koenigliche Technische Hochschule zu Muenchen, 1891/92 to 1913/14.

Chronik der Preussischen vereinigten Friedrichs-Universitaet Halle-Wittenberg (Halle: Buchdruckerei des Waisenhauses) 1901–02, 1916–1926.

Deutsche Hochschulstatistik (Berlin: Struppe und Winckler), 1928.

Deutsche Landwirtschaftliche Presse, 1918–1930.

Illustrierte Landwirtschaftliche Zeitung, 1918–1930.

Jahresberichte der Wuerttembergischen Landwirtschaftlichen Hochschule Hohenheim, 1900–1914.

[Die Koenigliche] Bayerische Hochschule [Akademie] fuer Landwirtschaft und Brauerei Weihenstephan (Freising: Datterer) 1898, 1925, 1929.

Koenigliche Landwirtschaftliche Akademie Bonn-Poppelsdorf: Jahresberichte, 1909–1920.

Landwirtschaftliche Hochschule Berlin: Jahresberichte, 1892–1921.

Landwirtschaftliche Jahrbuecher, supplementary volumes, 1903 and 1910.

Mentzel und von Lengerkes Landwirtschaftlicher Huelfs- und Schreibkalender (Berlin: Parey), 1902–1930.

Mitteilungen der Deutschen Landwirtschafts-Gesellschaft, 1919–1930.

Mitteilungen der landwirtschaftlichen Institute der Koeniglichen Universitaet Breslau, 1899–1914.

Satzungen der [Koeniglichen] Landwirtschaftlichen Akademie [Hochschule] Bonn (Bonn: Georgi) 1900, 1920.

Statistisches Jahrbuch fuer das Deutsche Reich (Berlin: Puttkammer and Muehlbrecht).

Uebersicht ueber die Organisation, die Zwecke, den Lehrplan [...] der Koeniglichen Wuerttembergischen Landwirtschaftlichen Akademie Hohenheim (Plieningen: Find) 1899, 1901, 1904, 1906, 1909, 1911.

Wochenblatt des Landwirtschaftlichen Vereins in Bayern, 1918–1920.

Wuerttembergisches Wochenblatt fuer Landwirtschaft.

Books and Articles

Abderhalden, E. 'Ein Vorstoss gegen die landwirtschaftliche Institute, insbes. gegen dasjenige der Universitaet Halle.' *Hallische Universitaets-Zeitung*, 1 December 1919, 676–678.

Achilles, Walter. *Deutsche Agrargeschichte im Zeitalter der Reformen und der Industrialisierung*. Stuttgart: Ulmer, 1993.

Aereboe, Friedrich. 'Landwirt. Hochschulen oder landwirt. Universitaets-institute?' *Deutsche Landwirtschaftliche Presse*, 29 October 1919, 659–660.

——. *Allgemeine landwirtschaftliche Betriebslehre*. 5th ed. Berlin: Parey, 1920.

——. *Agrarpolitik: ein Lehrbuch*. Berlin: Parey, 1928.

——. 'Landwirtschaftliche Hochschulen.' In *Das Akademische Deutschland*, edited by Michael Doeberl et al., 633–644. Berlin: C.A. Weller, 1930.

——. *Wirtschaft und Kultur in den Vereinigten Staaten von Nordamerika*. Berlin: Parey, 1930.

——. *Jugendlust, Arbeitsfreuden und Arbeitskaempfe. Erinnerungen aus dem Leben von Fr. Aereboe.* unpublished typescript, n.d. (prob. ca. 1932).

Albrecht, Helmuth. *Technische Bildung zwischen Wissenschaft und Praxis: die Technische Hochschule Braunschweig, 1862–1914*. Hildesheim: Olms Weidmann, 1987.

Altrock, W. von. 'Landwirtschaftliches Vereinswesen.' In *Handwoerterbuch der Staatswissenschaften*, edited by Ludwig Elster et al., 212–218. Jena: G. Fischer, 1925.

Altrock, W. von. 'Landwirtschaftskammern.' In *Handwoerterbuch der Staatswissenschaften*, edited by Ludwig Elster et al., 220–229. Jena: Gustav Fischer, 1925.

Anderson, Robert. 'The origins of the International Rice Research Institute.' *Minerva* 29 (1991): 61–89.

Anderson, Robert, Edwin Levy and Barrie Morrison. *Rice Science and Development Politics: Research Strategies and IRRI's Technologies Confront Asian Diversity (1950–1980)*. Oxford: Clarendon Press, 1991.

Andreae, Bernd. 'Theodor Brinkmann und die landwirtschaftliche Betriebswissenschaft.' *Zeitschrift fuer Agrargeschichte und Agrarsoziologie* 8 (1960): 162–176.

Andreae, Friedrich. 'Schlesische Friedrich-Wilhelms-Universitaet Breslau.' In *Das Akademische Deutschland*, edited by Michael Doeberl et al., 97–110. Berlin: C.A. Weller Verlag, 1930.

Anon. *Die Koenigliche Preussische Landwirtschaftliche Akademie Bonn-Poppelsdorf in ihrer Entwicklung, von 1847–1900*. Bonn: Georgi, 1900.

——. 'Jahresbericht des Bayerischen Landwirtschaftsrates fuer 1907.' *Vierteljahresschrift des Bayerischen Landwirtschaftsrates* 13 (1908).

——. 'Ueber die Zuechtungsversuche zur Verbesserung des Landweizens Elsass-Lothringens.' *Fuehlings Landwirtschaftliche Zeitung* 60 (1911): 211–212.

——. *Das Studium der Landwirtschaft an der Koeniglich Wuerttembergischen Landwirtschaftlichen Hochschule Hohenheim*. Plieningen: Find, 1913.

——. 'Bericht der Koeniglich Bayerischen Saatzuchtanstalt fuer 1912 und 1913.' *Landwirtschaftliches Jahrbuch fuer Bayern* (1913).

——. *Koenigliche Landwirt. Akademie Bonn-Poppelsdorf: Ratschlaege fuer die Studieneinteilung der studierenden Landwirte*. Bonn: Georgi, 1914.

——. *Chronik der Rheinischen Friedrich-Wilhelms-Universitaet zu Bonn*. Vol. 44. Bonn, 1918.

——. [obituary of F. Wohltmann]. *Zeitschrift fuer Pflanzenzuechtung* 7 (1919): 77–78.

——. *Pruefungsordnung fuer Studierende der Landwirtschaft an Landwirtschaftlichen Hochschulen und mit landwirtschaftlichen Instituten ausgestatteten Universitaeten*. Berlin, 1922.

——. 'Gesellschaft der Foerderer der Landwirtschaftlichen Hochschule Berlin.' *Illustrierte Landwirtschaftliche Zeitung*, 20 January 1928, 35–36.

——, ed. *Kuerschners Deutscher Gelehrten-Lexikon*. Berlin: W. de Gruyter, 1931.

——, ed. *Geheimrat Th. Remy; +30. Dezember 1946; In memoriam*. Bonn: Rektor der Universitaet Bonn, 1947.

——. 'The worst is still to come.' *Nature* (1985): 663.

Ash, Mitchell. *Gestalt Psychology in German Culture, 1890–1967: Holism and the Quest for Objectivity*. Cambridge: Cambridge University Press, 1995.

Ausschuss zur Untersuchung der Erzeugungs- und Absatzbedingungen der deutschen Wirtschaft. 'Das landwirtschaftliche Bildungs- und Beratungswesen in Deutschland.' Berlin, 1930.

Austoker, Joan. 'Walter Morley Fletcher and the origins of a basic biomedical research policy.' In *Historical Perspectives on the Role of the*

MRC, edited by Joan Austoker and Linda Bryder, 23–34. Oxford/New York: Oxford University Press, 1989.

Baumann, Hans. 'Acker- und Pflanzenbau.' In *Berliner Forschung und Lehre in den Landwirtschaftswissenschaften. Festschrift zur 75-Jahr-Feier der Landwirtschaftlich-Gaertnerischen Fakultaet der Humboldt-Universitaet*, edited by Bruno Skibbe and Henry Gocht, 113–121. Berlin: Humboldt-Universitaet zu Berlin, 1956.

Baumgarten, Marita. *Vom Gelehrten zum Wissenschaftler. Studien zum Lehrkoerper einer kleinen Universitaet am Beispiel der Ludoviciana Giessen (1815–1914)*. Giessen: Universitaetsbibliothek, 1988.

——. *Professoren und Universitaeten im 19. Jahrhundert*. Goettingen: Vandenhoeck und Ruprecht, 1997.

Baur, Erwin. 'Die Zukunft der preussischen landwirtschaftlichen Unterrichts- und Forschungsanstalten.' *Deutsche Landwirtschaftliche Presse*, 19 April 1919, 223–224.

——. *Die wissenschaftlichen Grundlagen der Pflanzenzuechtung: ein Lehrbuch fuer Landwirte, Gaertner und Forstleute*. 1st ed. Berlin: Borntraeger, 1921.

——. 'Zukunftsaufgaben und -Wege der Weizenzuechtung in Deutschland.' *Illustrierte Landwirtschaftliche Zeitung*, 8 January 1926, 17–18.

——. [opening address to 5th International Congress of Genetics 1927]. *Zeitschrift fuer induktive Abstammungs- und Vererbungslehre* suppl. Vol. I (1928): 51–53.

Becker, C.H. *Gedanken zur Hochschulreform*. Leipzig: Quelle und Meyer, 1919.

Becker, Heinrich. 'Von der Nahrungssicherung zu Kolonialtraeumen: die landwirtschaftlichen Institute im Dritten Reich.' In *Die Universitaet Goettingen unter dem Nationalsozialismus: das verdraengte Kapitel ihrer 250-jaehrigen Geschichte*, edited by Heinrich Becker, Hans-Joachim Dahms and Cornelia Wegeler, 410–436. Munich: K.G. Saur, 1987.

Becker, Heinrich and Guenther Schmitt. 'Die Entwicklung der Agrarwissenschaften an der Georg-August-Universitaet zu Goettingen.' *Zeitschrift fuer Agrargeschichte und Agrarsoziologie* 38 (1990): 42–61.

Behrend von Grass, Gerhard. 'Erinnerungen an Hohenheim, 1891–1912.' *Mitteilungen des Hochschulbundes und der Landwirtschaftlichen Hochschule Hohenheim* 1965, 50–56.

Benecke, Otto. *Die Pruefungen fuer Landwirte an den Universitaeten und Landwirtschaftlichen Hochschulen in Preussen*. Berlin: Weidmannsche Buchhandlung, 1925.

Bergmann, Hannsjoerg. *Der Bayerische Bauernbund und der Bayerische Christliche Bauernverein, 1919–1928*. Munich: C.H. Beck, 1986.

Biederbeck, A. 'Die Tagung der landwirt. Hochschulen und Universitaets-Institute in Halle.' *Deutsche Landwirtschaftliche Presse*, 3 July 1920, 381.

Blackbourn, David. 'Peasants and politics in Germany, 1871–1914.' *European History Quarterly* 14 (1984): 47–75.

Bode, Hugo. 'Vom Geist Julius Kuehns.' *Kuehn-Archiv* 3 (1913): 117–127.

Boehm, Wolfgang. *Einfuehrung in die Wissenschaftsgeschichte des Pflanzenbaus*. Goettingen: Triade-Verlag E. Claupein, 1990.

____. 'Geschichte des Landwirtschaftsstudiums in Deutschland.' *Berichte ueber Landwirtschaft* 66 (1988): 1–36.

____. *Biographisches Handbuch zur Geschichte des Pflanzenbaus*. Munich: K.G. Saur, 1997.

Boehmer, Dr. 'Zur Frage des hoeheren landwirtschaftlichen Unterrichts.' *Deutsche Landwirtschaftliche Presse*, 3 December 1919, 739.

Boelcke, W.A. 'Die Gruendungsgeschichte der Hohenheimer Ordinariate 1818 bis 1945.' *Jahrbuch fuer Statistik und Landeskunde von Baden-Wuerttemberg* 8 (1964): 96–114.

Bonner, Thomas. *Becoming a Physician: Medical Education in Great Britain, France, Germany and the United States, 1750–1945*. New York: Oxford University Press, 1995.

Borscheid, Peter. *Naturwissenschaft, Staat und Industrie in Baden (1848–1914)*. Stuttgart: Ernst Klett Verlag, 1976.

Bourdieu, Pierre. 'The specificity of the scientific field and the social conditions of the progress of reason.' *Social Science Information* 14 (1975): 19–47.

——. *Homo Academicus*. Stanford, CA: Stanford University Press, 1988.

——. 'The field of cultural production or: the economic world reversed.' In *The Field of Cultural Production: Essays on Art and Literature*, edited by Pierre Bourdieu, 29–73. Cambridge: Polity Press, 1993.

Braun, Hans-Joachim. 'Methodenprobleme der Ingenieurwissenschaft, 1850 bis 1900.' *Technikgeschichte* 44 (1977): 1–18.

Bray, Francesca. *Technology and Gender: Fabrics of Power in Late Imperial China*. Berkeley/London: University of California Press, 1997.

Brierley, William. 'Some viewpoints of an applied biologist.' *Annals of Applied Biology* 21 (1934): 351–378.

Brinkmann, Theodor. 'Die Landwirtschaftliche Betriebslehre in Bonn-Poppelsdorf.' In *Bonner Gelehrte. Beitraege zur Geschichte der*

Wissenschaften in Bonn, 21–32. Bonn: Bouvier Verlag and Roehrscheid Verlag, 1971.
Brittain, James and Robert McMath. 'Engineers and the New South Creed: the Formation and Early Development of Georgia Tech.' In *The Engineer in America: a Historical Anthology from Technology and Culture*, edited by Terry S. Reynolds, 123–149. Chicago: Chicago University Press, 1991.
Brocke, Bernhard vom. 'Friedrich Althoff: a great figure in higher education policy in Germany.' *Minerva* 29 (1991): 269–293.
Broman, Thomas. 'Rethinking professionalization: theory, practice and professional ideology in 18th century German medicine.' *Journal of Modern History* 67 (1995): 835–872.
Brose, Eric D. *The Politics of Technological Change in Prussia, 1809–1848: out of the Shadow of Antiquity*. Princeton, NJ: Princeton University Press, 1993.
Bruch, Ruediger vom. 'Von der Sozialethik zur Sozialtechnologie? Neuorientierungen in der deutschen Sozialwissenschaft um 1900.' In *Kultur und Kulturwissenschaften um 1900, Band 2: Idealismus und Positivismus*, edited by Gangolf Huebinger, Ruediger vom Bruch and Friedrich Wilhelm Graf, 260–276. Stuttgart: Franz Steiner, 1997.
Buelow, Friedrich. 'Grundlagen, Entwicklung und Bedeutung der Landwirtschaftlichen Fakultaet an der Friedrich-Wilhelms Universitaet Berlin bis zum Jahr 1945.' In *Studium Berolinense*, edited by H. Leussing et al., 905–919. Berlin: Walter de Gruyter, 1960.
Busch, Lawrence and William Lacey. *Science, Agriculture, and the Politics of Research*. Boulder/London: Westview Press, 1983.
——, eds. *The Agricultural Scientific Enterprise: a System in Transition*. Boulder/London: Westview Press, 1986.
Busch, Lawrence et al. *Plants, Power and Profit: Social, Economic and Ethical Consequences of the New Biotechnologies*. Cambridge, MA/-Oxford: Blackwell, 1991.
Buttel, Frederick and Lawrence Busch. 'The public agricultural research system at the crossroads.' *Agricultural History* 62 (1988): 303–324.
Buttel, Frederick et al. 'Industry/land grant university relationships in transition.' In *The Agricultural Scientific Enterprise: a System in Transition*, edited by Lawrence Busch and William Lacey, 296–314. Boulder/London: Westview Press, 1986.
Calvert, Monte. *The Mechanical Engineer in America, 1830–1910*. Baltimore: Johns Hopkins University Press, 1967.

Carmon, Arye. 'The diverse and changing fortunes of the University of Heidelberg under National Socialism.' *Minerva* 16 (1978): 516–544.

Channell, David. 'The harmony of theory and practice: the engineering science of W.J.M. Rankine.' *Technology and Culture* 23 (1982): 39–52.

Churchill, Frederick. 'Wilhelm Roux.' In *Dictionary of Scientific Biography*, edited by C.C. Gillispie, 570–575. New York: Scribners, 1975.

Ciesla, Burghard. 'Abschied von der "reinen" Wissenschaft: "Wehrtechnik" und Anwendungsforschung in der Preussischen Akademie nach 1933.' In *Die Preussische Akademie der Wissenschaften zu Berlin 1914–1945*, edited by W. Fischer, Rainer Hohlfeld and Peter Noetzoldt, 483–513. Berlin: Akademie Verlag, 2000.

Comberg, Gustav et al. *Die deutsche Tierzucht im 19. und 20. Jahrhundert*. Stuttgart: Verlag Eugen Ulmer, 1984.

Committee of Inquiry into the Engineering Profession. *Engineering our Future*. London: HMSO, 1980.

Conrad, J. *Lebenserinnerungen*. unpublished manuscript, Staatsbibliothek Preussischer Kulturbesitz, Berlin, 1917.

Cooke, Kathy. 'From science to practice or practice to science? chickens and eggs in Raymond Pearl's agricultural breeding research, 1907–1916.' *Isis* 88 (1997): 62–86.

Danbom, David. *The Resisted Revolution: Urban America and the Industrialization of Agriculture, 1900–1930*. Ames: Iowa State University Press, 1979.

——. 'Publicly sponsored agricultural research in the United States from an historical perspective.' In *New Directions for Agriculture*, edited by K. Dahlberg. Totowa, NJ: Rowman and Allanheld, 1986.

Day, Charles R. *Education for the Industrial World: the Ecole d'Arts et Metiers and the Rise of French Industrial Engineering*. Cambridge, MA: MIT Press, 1987.

DeJager, Timothy. 'Pure science and practical interests: the origin of the Agricultural Research Council, 1930–1937.' *Minerva* 31 (1993): 129–150.

Delbrueck, Max. *Die Koenigliche Landwirtschaftliche Hochschule in der Zukunft*. Berlin: Parey, 1900.

Dienel, Hans-Liudger. 'Der Muenchener Weg im Theorie-Praxis-Streit um die Emanzipation des Wissenschaftlichen Maschinenbaus.' In *Die Technische Universitaet Muenchen: Annaeherungen an ihre Geschichte*, edited by Ulrich Wengenroth, 87–115. Munich: Technische Universitaet Muenchen, 1993.

——. *Ingenieure zwischen Hochschule und Industrie: Kaeltetechnik in Deutschland und Amerika, 1870–1930*. Goettingen: Vandenhoeck und Ruprecht, 1995.

Dienel, Hans-Liudger and Helmut Hilz, ed. *125 Jahre Technische Universitaet Muenchen, 1868–1993. Bayerns Weg in das Technische Zeitalter*. Munich: Hugendubel, 1993.

Diener, Hans Oskar. 'Zur Geschichte von Weihenstephan. Ein Beitrag zur Geschichte der Landwirtschafts-Lehre in Bayern.' *Zeitschrift fuer bayerische Landesgeschichte* 1 (1928): 453–476.

——. 'Paul Kulisch. Der erste und einzige Wahlrektor der Hochschule fuer Landwirtschaft und Brauerei in Weihenstephan.' *Bayerisches landwirtschaftliches Jahrbuch* 51 (1974): 717–767.

Disselhorst, R. 'Gedaechtnisrede auf F. Wohltmann.' *Hallische Universitaets-Zeitung* 1919, 210–214.

Dittrich, M. 'Julius Kuehn und sein Landwirtschaftliches Institut in Halle: ein Beitrag zur Geschichte der wissenschaftlichen Methodik.' *Zeitschrift fuer Agrargeschichte und Agrarsoziologie* 10 (1962): 61–70.

Divall, Colin. 'Fundamental science versus design: employers and engineering studies in British universities, 1935–1976.' *Minerva* 29 (1991): 167–194.

Dolezal, E. 'Hofrat Prof.Dr. agrar., Dr. agrar. h.c. Carl Fruwirth. Ein Lebensbild.' *Oesterreichische Zeitschrift fuer Vermessungswesen* (1931): 1–8.

Duewell, Kurt. 'Die Neugestaltung der Technischen Hochschulen nach dem ersten Weltkrieg.' *Technikgeschichte* 36 (1969): 220–244.

Ede, R. 'The School of Agriculture, University of Cambridge.' *Agricultural Progress* 15 (1938): 137–142.

Elleke, [?]. 'Noch einmal: Zur Frage der Beseitigung der landwirtschaftlichen Universitaetsinstitute.' *Deutsche Landwirtschaftliche Presse*, 25 October 1919, 653–654.

Endres, R. 'Zur Fruehgeschichte des landwirtschaftlichen Bildungswesens in Deutschland.' In *Wirtschaftskraefte und Wirtschaftswege: Festschrift fuer H. Kellenbenz*, edited by J. Schneider, 187–202. Stuttgart: Klett-Cotta, 1978.

Engelbrecht, Th. H. *Die Landbauzonen der aussertropischen Laender.* Vol. 1. Berlin: Dietrich Reimer, 1899.

Farr, Ian. 'Populism in the countryside: the Peasant Leagues in Bavaria in the 1890s.' In *Society and Politics in Wilhelminian Germany*, edited by R.J. Evans, 136–159. London: Croom Helm, 1978.

——. 'Peasant protest in the Empire: the Bavarian example.' In *Peasants and Lords in Modern Germany*, edited by Robert G. Moeller, 110–139. Boston: Allen and Unwin, 1986.

Feige, Ernst. 'Die gegenwaertige Gestaltung des landwirtschaftlichen Hochschulunterrichts in Deutschland.' *Die Naturwissenschaften* 3 October (1913): 958–960.

——. 'Die wissenschaftliche Stellung und Aufgabe der Landwirtschaft.' *Die Naturwissenschaften* 2 May (1913): 429–433.

Ferber, Christian von. *Die Entwicklung des Lehrkoerpers der deutschen Universitaeten und Hochschulen 1864–1954*. Goettingen: Vandenhoeck und Ruprecht, 1956.

Ferguson, Eugene. *Engineering and the Mind's Eye*. Cambridge, MA: MIT Press, 1992.

Feulner. 'Landwirtschaftliche Berufsbildung.' *Wochenblatt des Landwirtschafts-Vereins in Bayern* 110 (1920): 204–205.

Finlay, Mark. 'Science, Practice and Politics: German Agricultural Experiment Stations in the 19th Century.' PhD, Iowa State University, 1992.

——. 'Death of the academies: practice and science in German agricultural education.' Paper presented at the Anglo-North American conference in History of Science, Toronto 1992.

——. 'Theory versus practice in agricultural science: agricultural education and the German nation in the later 19th century.' (unpublished typescript) (1995).

Fitzgerald, Deborah. *The Business of Breeding: Hybrid Corn in Illinois, 1890–1940*. Ithaca, NY: Cornell University Press, 1990.

——. *Every Farm a Factory: the Industrial Ideal in American Agriculture*. New Haven/London: Yale University Press, 2003.

Flaig, W. 'Eine ausserordentliche Vertreterversammlung aller landwirtschaftlichen Hochschulen [...]' *Mitteilungen der Deutschen Landwirtschafts-Gesellschaft*, 26 June 1920, 355–356.

Flaig, W. and Hans Zoerner. 'Nochmals: "Die Tagung der landwirt. Hochschulen und Universitaets-Institute in Halle."' *Deutsche Landwirtschaftliche Presse*, 14 July 1920, 398.

Fletcher, T.W. 'The great depression of English agriculture, 1873–1896.' In *British Agriculture, 1875–1914*, edited by P.J. Perry, 30–55. London: Methuen, 1973.

Franz, Guenther. *Liebig und Hohenheim*. (vol. 15 of *Landwirtschaftliche Hochschule Hohenheim. Reden und Abhandlungen)*. Stuttgart: Eugen Ulmer, 1963.

——, ed. *Universitaet Hohenheim: landwirtschaftliche Hochschule, 1818–1968*. Stuttgart: Eugen Ulmer, 1968.
——. 'Die Geschichte der Universitaet.' In *Universitaet Hohenheim: Landwirtschaftliche Hochschule, 1818–1968*, edited by Guenther Franz, 11–160. Stuttgart: Ulmer, 1968.
Franz, Guenther and Heinz Haushofer, ed. *Grosse Landwirte*. Frankfurt a.M.: DLG-Verlag, 1970.
Franz, Heike. 'Betriebswirte in Deutschland 1900–1930: "Buerger" oder "Professionals"? Selbstbild und Fremdbild einer "neuen" akademischen Berufsgruppe.' In *Wege zur Geschichte des Buergertums*, edited by Klaus Tenfelde and Hans-Ulrich Wehler, 249–272. Goettingen: Vandenhoeck und Ruprecht, 1994.
——. *Zwischen Markt und Profession: Betriebswirte in Deutschland im Spannungsfeld von Bildungs- und Wirtschaftsburgertum (1900–1945)*. Goettingen: Vandenhoeck and Ruprecht, 1998.
Frauendorfer, Sigmund von. *Ideengeschichte der Agrarwirtschaft und Agrarpolitik im deutschen Sprachgebiet. Teil 1 – von den Anfaengen bis zum ersten Weltkrieg*. Munich/Bonn/Vienna: Bayerischer Landwirtschafts-Verlag, 1957.
Friedman, Robert Marc. *Appropriating the Weather: Vilhelm Bjerknes and the Construction of a Modern Meteorology*. Ithaca, NY: Cornell University Press, 1989.
Frimmel, F. 'Hofrat Prof. Dr. Dr. h.c. E. Tschermak zu seinem 80. Geburtstag.' *Zeitschrift fuer Pflanzenzuechtung* 30 (1951): 345–349.
Fritsch, Werner and Herbert Gottwald. 'Bayerische Bauernbund, 1895–1933.' In *Lexikon zur Parteiengeschichte. Die buergerlichen und kleinbuergerlicher Parteien und Verbaende in Deutschland (1789–1945)*, edited by Dieter Fricke, 131–151. Leipzig: VEB Bibliographisches Institut, 1983.
Froelich, Gustav. 'Nochmals: Landwirtschaftliche Hochschulen oder landwirtschaftliche Universitaetsinstitute?' *Deutsche Landwirtschaftliche Presse*, 19 November 1919, 708–709.
——. 'Das Tierzucht-Versuchsgut Domaene Lettin bei Halle an der Saale.' In *Arbeiten aus den Landwirtschaftlichen Instituten der Universitaet Halle*, edited by Gustav Froelich, K. Steinbrueck and Theodor Roemer, 1-14. Berlin: Parey, 1926.
Fruwirth, Carl. *Die Zuechtung der landwirtschaftlichen Kulturpflanzen, vol. 1 – Allgemeine Zuechtungslehre*. Berlin: Parey, 1901.
——. *Wie kann sich der Landwirt Pflanzenzucht, Sortenversuche und Saatgutbau zu Nutze machen?* Berlin: Parey, 1906.

——. 'Formen des Pflanzenzuechtungs-betriebes.' *Fuehlings Landwirtschaftliche Zeitung* 59 (1910): 641–651.

——. 'Ueber den Unterricht in der landwirtschaftlichen Pflanzenzucht.' *Land- und Forstwissenschaftliche Unterrichtszeitung* 26 (1912): 1–27.

——. 'Zur Technik der Graszuechtung.' *Beitraege zur Pflanzenzucht* 3 (1913): 99–133.

Gerken, Julius. 'Die Tagung der Fachgruppen der landwirtschaftlichen Hochschulen und Universitaets-Institute in Halle.' *Niedersaechsische Hochschulzeitung*, 25 June 1920, 8.

Gerland, Dr. 'Nochmals: Landwirtschaftliche Hochschulen oder landwirtschaftliche Universitaetsinstitute?' *Deutsche Landwirtschaftliche Presse*, 15 November 1919, 701–702.

Giessmann, E.-J. 'Der Einfluss der Entwicklung der Natur- und Technikwissenschaften auf die Ausbildung landwirtschaftlicher Fachkraefte im 19. Jahrhundert.' *Dresdener Beitraege zur Geschichte der Technikwissenschaften* 17 (1989): 56–79.

Gillispie, C.C. 'The *Encyclopedie* and the Jacobin philosophy of science: a study in ideas and consequences.' In *Critical Problems in the History of Science*, edited by Marshall Clagett, 255–289. Madison, WI: Wisconsin University Press, 1959.

Gispen, Kees. *New Profession, Old Order: Engineers and German Society, 1815–1914*. Cambridge: Cambridge University Press, 1989.

Goltz, Theodor von der. *Geschichte der Deutschen Landwirtschaft*. Vol. 2. Stuttgart/Berlin: J.G. Cotta, 1903.

Goltz, Theodor von der, Otto Koll and Franz Kuenzel, eds. *Festschrift zur Feier des 50-jaehrigen Bestehens der Koeniglichen Preussischen Landwirtschaftlichen Akademie Poppelsdorf*. Bonn, 1897.

Graebke, Paul. 'Die Frage des Hochschul- oder Universitaetsstudiums der Landwirte im Lichte der wirtschaftlichen und kulturellen Entwicklung Deutschlands waehrend der letzten 50 Jahre vor dem Kriege.' *Deutsche Landwirtschaftliche Presse*, 10 and 17 April 1920, 213–214 and 222–223.

Grelon, Andre. 'The training and career structures of engineers in France, 1880–1939.' In *Education, Technology and Industrial Performance in Europe, 1850–1939*, edited by Robert Fox and Anna Guagnini, 42–64. Cambridge: Cambridge University Press, 1993.

——. 'Die deutschen Ingenieure aus franzoesischer Sicht, 1770–1990.' In *Ingenieure in Deutschland, 1770–1990*, edited by Peter Lundgreen and Andre Grelon, 369–386. Frankfurt/New York: Campus, 1994.

Gundler, Bettina. *Technische Bildung, Hochschule, Staat und Wirtschaft: Entwicklungslinien des Technischen Hochschulwesens 1914 bis 1930. das Beispiel der TH Braunschweig*. Hildesheim: Olms Weidmann, 1992.

Hacker, Sally. 'Mathematisation of engineering: limits on women and the field.' In *Machina Ex Dea: Feminist Perspectives on Technology*, edited by Joan Rothschild, 38–58. New York/Oxford: Pergamon Press, 1983.

Hadlington, Simon. 'Unilever purchases British plant breeding interests.' *Nature* 328 (1987): 569.

Hadwiger, Don. *The Politics of Agricultural Research*. Lincoln, NE and London: University of Nebraska Press, 1982.

Haenseroth, Thomas and Klaus Mauersberger. 'Das Dresdener Konzept zur Genese technikwissenschaftlicher Disziplinen – eine Bilanz.' *Dresdener Beitraege zur Geschichte der Technikwissenschaften* 24 (1996): 20–45.

Hansen, J. *Das landwirtschaftliche Unterrichtswesen und die Ausbildung des Landwirts*. Berlin: Parey, 1919 (2nd ed. 1920).

——. 'Zur Ausbildung des Landwirts.' *Deutsche Landwirtschaftliche Presse*, 13 September 1919, 562.

——. 'Landwirtschaftliches Universitaets-Institut oder landwirtschaftliche Hochschule.' *Illustrierte Landwirtschaftliche Zeitung*, 15 November 1919, 454–455.

Hard, Mikael. *Machines are Frozen Spirit: the Scientification of Refrigeration and Brewing in the 19th Century – a Weberian Interpretation*. Frankfurt a.M.: Campus Verlag and Boulder, CO: Westview Press, 1994.

Harlan, H.V. *One Man's Life with Barley: the Memories and Observations of Harry V. Harlan*. New York, 1957.

Harms, H. 'Ludwig Wittmack.' *Berichte der Deutschen Botanischen Gesellschaft* 49 (1931): (200)–(219).

Harwood, Jonathan. *Styles of Scientific Thought: the German Genetics Community, 1900–1933*. Chicago: University of Chicago Press, 1993.

____. 'Institutional Innovation in fin de siecle Germany.' *British Journal for History of Science* 27 (1994): 197–211.

——. 'The reception of genetic theory among academic plant-breeders in Germany, 1900–1930.' *Journal of the Swedish Seed Association* 107 (1997): 187–195.

——. 'The rediscovery of Mendelism in agricultural context: Erich von Tschermak as plant-breeder.' *Comptes rendus de l'Academie des*

Sciences: Sciences de la Vie/Life Sciences 323, no. 12 (2000): 1061–1067.

——. 'Politische Oekonomie der Pflanzenzucht in Deutschland, ca. 1870–1933.' In *Autarkie und Ostexpansion: Pflanzenzucht und Agrarforschung im Nationalsozialismus*, edited by Susanne Heim. Goettingen: Wallstein, 2002.

Hashagen, Ulf. 'Der Mathematiker Walther von Dyck und die "wissenschaftliche" Technische Hochschule.' In *Oszillationen: Naturwissenschaftler und Ingenieure zwischen Forschung und Markt*, edited by Ivo Schneider, Helmuth Trischler, and Ulrich Wengenroth, 267–296. Munich/Vienna: R. Oldenbourg, 2000.

Haushofer, Heinz. *Die Furche der Deutschen Landwirtschafts-Gesellschaft, 1885–1960*. Frankfurt a.M.: DLG-Verlag, 1960.

——. *Ein halbes Jahrhundert im Dienste der bayerischen Landwirtschaft und der Volksernaehrung – Zur Geschichte des Bayerischen Staatsministerium fuer Ernaehrung, Landwirtschaft und Forsten*. Munich: BLV Verlagsgesellschaft, 1969.

——. 'Bauer im Schach. Lebenserinnerungen.' *Bayerisches landwirtschaftliches Jahrbuch* 59, special issue (1982): 1–236.

——. 'Baeuerliche Fuehrungsschichten in Bayern im 19.–20. Jahrhundert.' In *Aus der Bayerischen Agrargeschichte, 1525–1978. Gesammelte Beitraege zur Bayerischen Agrargeschichte von Heinz Haushofer*, edited by Pankraz Fried and Wolfgang Zorn, 39–48. Munich/Vienna/Zurich: BLV Verlagsgesellschaft, 1986.

Heape, Walter. *The Breeding Industry: its Value to the Country and its Needs*. Cambridge: Cambridge University Press, 1906.

Heim, Susanne. *Kalorien, Kautschuk, Karrieren: Pflanzenzuechtung und landwirtschaftliche Forschung in Kaiser-Wilhelm-Instituten 1933–1945*. Goettingen: Wallstein, 2003.

Hensel, Susann. 'Die Auseinandersetzungen um die mathematische Ausbildung der Ingenieure an den Technischen Hochschulen in Deutschland Ende des 19. Jahrhunderts.' In *Mathematik und Technik im 19. Jahrhundert in Deutschland*, edited by S. Hensel, K.-N. Ihmig and M. Otte, 1–111. Goettingen: Vandenhoeck und Ruprecht, 1989.

Henseler, H. 'Die landwirtschaftliche Abteilung der T.H. Muenchen.' *Wochenblatt des Landwirtschaftlichen Vereins in Bayern* 1929, 410–412.

Herbst, H.-G. 'Zur Geschichte der Zoologie an den landwirtschaftlichen Hochschuleinrichtungen in Berlin.' *Wissenschaftliche Zeitschrift der*

Humboldt-Universitaet zu Berlin, Mathe.-Naturw. Reihe 34 (1985): 330–341.
Herpel, Hans Joachim. *Die Entwicklung des landwirtschaftlichen Studiums an der Universitaet Goettingen.* Goettingen: Vandenhoeck und Ruprecht, 1932.
Heumann, Landesoekonomierat and Rittergutsbesitzer von Campe. 'Landwirtschaftliche Fachhochschulen oder landwirtschaftliche Abteilungen an den Universitaeten.' *Illustrierte Landwirtschaftliche Zeitung*, 4 October 1919, 402.
Hightower, Jim. *Hard Tomatoes, Hard Times.* Cambridge, MA: Schenkman, 1973.
Hillmann, Paul. 'Das wissenschaftliche Studium der Landwirtschaft.' *Fuehlings Landwirtschaftliche Zeitung* 69 (1920): 121–130.
Hilz, Helmut. 'Weihenstephan: Landwirte- und Brauereiausbildung mit Tradition.' In *Die Technische Universitaet Muenchen. Annaeherungen an ihre Geschichte*, edited by Ulrich Wengenroth, 143-155. Munich: Technische Universitaet Muenchen, 1993.
Hochberger, Anton. *Der Bayerische Bauernbund, 1893–1914.* Munich: C.H. Beck, 1991.
Hofstadter, Richard. *Anti-Intellectualism in American Life.* New York: Vintage Books, 1962.
Holdefleiss, F. 'Das landwirtschaftliche Studium.' In *Festschrift zur Feier des 100-jaehrigen Bestehens der Universitaet Breslau*, edited by G. Kaufmann, 504–516. Breslau: Hirt, 1911.
Holmes, C.J. 'Science and the farmer: the development of the agricultural advisory service in England and Wales, 1900–1939.' *Agricultural History Review* 36 (1988): 77–86.
Hoppe, Hans. 'Die Ausbildung des Landwirts.' *Fuehlings Landwirtschaftliche Zeitung* 70 (1921): 321–335.
Hortleder, Gerd. *Das Gesellschaftsbild des Ingenieurs: zum politischen Verhalten der technischen Intelligenz in Deutschland.* 2nd ed. Frankfurt a.M.: Suhrkamp, 1970.
Howarth, Janet. 'Science education in late Victorian Oxford: a curious case of failure?' *English Historical Review* 102 (1987): 334–371.
Huber, Ernst Rudolf. *Deutsche Verfassungsgeschichte seit 1789, Band 4 – Struktur und Krisen des Kaiserreichs.* Stuttgart: W. Kohlhammer, 1969.
Huettermann, Aloys. 'History of forest botany (Forstbotanik) in Germany from the beginning in 1800 until 1940 – science in the tension field between basic and applied sciences.' *Berichte der Deutschen Botanischen Gesellschaft* 100 (1987): 107–141.

Hughes, Thomas. *Networks of Power: Electrification in Western Society, 1880–1930*. Baltimore/London: Johns Hopkins University Press, 1983.

Hunt, Bruce. '"Practice versus theory": the British electrical debate, 1888–1891.' *Isis* 74 (1983): 341–355.

Hutchison, C.B. and A.R. Mann. 'Higher Education in Agriculture in Europe.' Paris: International Education Board, 1927.

Jacob, F.H. 'Reflections of an applied biologist.' *Annals of Applied Biology* 87 (1977): 295–306.

Jacyna, L.S. 'The lab and the clinic: the impact of pathology on surgical diagnosis in Glasgow Western Infirmary, 1875–1920.' *Bulletin of the History of Medicine* 62 (1988): 384–406.

Jansen, Sarah. *'Schaedlinge': Geschichte eines wissenschaftlichen und politischen Konstrukts 1840–1920*. Frankfurt a.M.: Campus Verlag, 2003.

Jarausch, Konrad. *The Unfree Professions: German Lawyers, Teachers and Engineers, 1900–1950*. New York/Oxford: Oxford University Press, 1990.

——. 'Universitaet und Hochschule.' In *Handbuch der deutschen Bildungsgeschichte*, edited by C. Berg, 313–345. Munich: C.H. Beck, 1993.

Jas, Nathalie. *Au Carrefour de la Chimie et de l'Agriculture: les sciences agronomiques en France et en Allemagne, 1840–1914*. Paris: Edition des archives contemporaines, 2001.

Jenne, Heinrich. *Das landwirtschaftliche Unterrichtswesen in Bayern*. Leipzig: A. Deichert, 1906.

Jennings, Bruce. *Foundations of International Agricultural Research: Science and Politics in Mexican Agriculture*. Boulder, CO: Westview Press, 1988.

Joravsky, David. *The Lysenko Affair*. Chicago: University of Chicago Press, 1970.

Kalt, Bertram. 'Ferdinand Wohltmann, sein Werden und Wirken als Hochschullehrer und Forscher.' *Hallische Universitaets-Zeitung* 1919, 214–239.

Kattenbusch, F. 'Zum Gedaechtnis F. Wohltmanns.' *Hallische Universitaets-Zeitung* 1919, 201–210.

Kaufman, Martin. 'American medical education.' In *The Education of American Physicians: Historical Essays*, edited by R.L. Numbers, 7–28. Berkeley, CA: University of California Press, 1980.

Kaufmann, G., ed. *Festschrift zur Feier des 100-jaehrigen Bestehens der Universitaet Breslau*. Breslau: Hirt, 1911.

Kenney, Martin. *Biotechnology: the University-Industrial Complex*. New Haven/London: Yale University Press, 1986.

Kerschowsky, Oekonomierat. 'Das landwirtschaftliche Unterrichtswesen und die Ausbildung des Landwirts.' *Deutsche Landwirtschaftliche Presse*, 21 June 1919, 374–375.

Kick, Hermann. 'Kurze Geschichte der Landwirtschaftlichen Fakultaet.' In *Bonner Gelehrte: Beitraege zur Geschichte der Wissenschaften in Bonn*, 10–20. Bonn: Bouvier, Roehrscheid, 1971.

Kiessling, Ludwig. 'Einiges aus der Praxis des Zuchtgartenbetriebes.' *Zeitschrift fuer Pflanzenzuechtung* 1 (1912–13): 25–36.

——. 'Die Stellung der Landwirtschaftslehre zu den technischen und den uebrigen Wissenschaften.' Published inaugural address, Munich, 10 December 1921.

——. 'Wissenschaftliche und praktische Pflanzenbaufoerderung.' *Illustrierte Landwirtschafts-Zeitung*, 17 June 1922, 203–204.

——. 'Einige Gegenwartsfragen der Pflanzenzuechtung und des Saatbaues.' *Mitteilungen der Deutschen Landwirtschafts-Gesellschaft* 1922, 454–460.

____. 'Zur Problemstellung, Begriffsbestimmung und Methodik der Pflanzenzucht.' *Beitraege zur Pflanzenzucht* 7 (1924): 11–21.

——. 'K. von Ruemker und die deutsche Pflanzenzuechtung.' In *Forschungen auf dem Gebiete des Pflanzenbaus und der Pflanzenzuechtung. Festschrift zum 70. Geburtstag von Kurt von Ruemker*, 1–9. Berlin: Parey, 1929.

Kiessling, Ludwig and A. Gassner. *Entwicklung und Taetigkeit der Bayerischen Landessaatzuchtanstalt im ersten Vierteljahr ihres Bestehens*. Munich: Carl Gerber, 1927.

Kimmelman, Barbara. 'A Progressive Era Discipline: Genetics in American Agricultural Colleges and Experiment Stations, 1900–1920.' PhD, University of Pennsylvania, 1987.

Kirchgaessner, Bernhard. 'Die Gruendung der Handelshochschulen Frankfurt und Mannheim als Leistung des Besitz- und Bildungsbuergertums.' In *Stadt und Hochschule im 19. und 20. Jahrhundert*, edited by Erich Maschke and Juergen Sydow, 123–139. Sigmaringen: Jan Thorbecke Verlag, 1979.

Kirchner, Oskar. *Die Entwicklung der Koeniglichen landwirtschaftlichen Anstalt Hohenheim*. Stuttgart: Ulmer, 1918.

Kirchner, Prof. Dr. 'Landwirthschaft und Universitaet.' *Fuehlings Landwirtschaftliche Zeitung* 50 (1901): 16–10.

Klapp, Ernst. 'Theodor Remy 1868–1946.' In *Bonner Gelehrte. Beitraege zur Geschichte der Wissenschaften in Bonn*, 116–136. Bonn: Bouvier Verlag, Roehrscheid Verlag, 1971.

Klein, Ernst. *Die akademischen Lehrer der Universitaet Hohenheim (Landwirtschaftliche Hochschule), 1818–1968*. Stuttgart: W. Kohlhammer, 1968.

Klein, H. 'Die Entwicklung der Agrarwissenschaften von der Gruendung der Landwirtschaftlichen Hochschule Berlin bis zu den agrarwissenschaftlichen Sektionen an der Humboldt-Universitaet zu Berlin.' In *Festschrift anlaesslich der 100. Wiederkehr des Gruendungsjahres der Landwirt. Hochschule Berlin und des 30jaehrigen Bestehens der Akademie der Landwirtschaftswissenschaften der DDR*, edited by Willi Breunig, Helmut Bochow and Dieter Spaar, 7–36. Berlin: Humboldt-Universitaet zu Berlin, 1981.

Kleinberger, Aharon. 'Gab es eine nationalsozialistische Hochschulpolitik?' In *Erziehung und Schulung im Dritten Reich. Teil 2: Hochschule, Erwachsenenbildung*, edited by Manfred Heinemann, 9–30. Stuttgart: Klett-Cotta, 1980.

Klemm, Volker. *Agrarwissenschaften in Deutschland: Geschichte – Tradition von den Anfaengen bis 1945*. St. Katharinen: Scripta Mercaturae Verlag, 1992.

——. *Agrarwissenschaften im 'Dritten Reich': Aufstieg oder Stuerz? (1933–1945)*. Berlin: Foerdergesell. A.D. Thaer and Landwirtschaftlich-Gaertnerische Fakultaet der Humboldt-Universitaet zu Berlin, 1994.

Klemm, Volker et al. *Von der Koeniglichen Akademie des Landbaus in Moeglin zur Landwirtschaftlich-Gaertnerischen Fakultaet der Humboldt-Universitaet zu Berlin*. Berlin: Humboldt-Universitaet zu Berlin, 1998.

Kline, Ronald. 'Construing "technology" as "applied science": public rhetoric of scientists and engineers in the United States, 1880–1945.' *Isis* 86 (1995): 194–221.

Kloppenburg, Jack. *First the Seed: the Political Economy of Plant Biotechnology, 1492–2000*. Cambridge: Cambridge University Press, 1988.

Kluge, Rittergutsbesitzer. 'Das Hochschulstudium fuer praktische Landwirte.' *Deutsche Landwirtschaftliche Presse*, 31 January 1920, 66.

Kocka, Juergen. *Die Angestellten in der deutschen Geschichte, 1850–1980: vom Privatbeamten zum angestellten Arbeitnehmer*. Goettingen: Vandenhoeck and Ruprecht, 1981.

Koenig, Wolfgang. 'Stand und Aufgaben der Forschung zur Geschichte der deutschen Polytechnischen Schulen und Technischen Hochschulen im 19. Jahrhundert.' *Technikgeschichte* 48 (1981): 47–67.

——. 'Spezialisierung und Bildungsanspruch: zur Geschichte der Technischen Hochschulen im 19. und 20. Jahrhundert.' *Berichte zur Wissenschaftsgeschichte* 11 (1988): 219–225.

——. 'Science-based industry or industry-based science: electrical engineering in Germany before World War One.' *Technology and Culture* 37 (1996): 70–101.

——. *Kuenstler und Strichezieher: Konstruktions- und Technikkulturen im deutschen, britischen, amerikanischen und franzoesischen Maschinenbau zwischen 1850 und 1930*. Frankfurt: Suhrkamp, 1999.

Koernicke, Max. 'Die Landwirtschaftliche Hochschule Bonn-Poppelsdorf.' *Bonner Mitteilungen* 7 (1931): 1–6.

Kohler, Robert. *Partners in Science: Foundations and Natural Scientists, 1900–1945*. Chicago/London: University of Chicago Press, 1991.

Kraemer, H. 'Das landwirtschaftliche Unterrichtswesen und die Ausbildung des Landwirts [review].' *Mitteilungen der Deutschen Landwirtschafts-Gesellschaft*, 27 September 1919, 524–526.

Kranakis, Eda. 'Hybrid careers and the interaction of science and technology.' In *Technological Development and Science in the Industrial Age*, edited by Peter Kroes and M. Bakker, 177–204. Dordrecht: Kluwer, 1992.

——. *Constructing a Bridge: an Exploration of Engineering Culture, Design and Research in 19th Century France and America*. Cambridge, MA: MIT Press, 1997.

Kraus, Carl. 'Die Pflege der Landwirtschaftswissenschaft in Bayern.' *Jahrbuch der T.H. Muenchen* (1903–04): 3–29.

Kryzymowski, Richard. 'Beziehungen zwischen der Betriebsintensitaet und der Sortenfrage.' *Jahrbuch der Deutschen Landwirtschafts-Gesellschaft* 28 (1913): 456–467.

____. *Philosophie der Landwirtschaftslehre*. Stuttgart: Eugen Ulmer, 1919.

____. *Geschichte der deutschen Landwirtschaft*. 2nd ed. Stuttgart: Eugen Ulmer, 1951.

Kuehn, Julius. *Programm fuer das Studium der Landwirtschaft an der Universitaet Halle*. Halle: Buchhandlung des Waisenhauses, 1905.

Kulisch, Paul. 'Die staatliche Foerderung der Saatzucht und des Saatgutbaues in Elsass-Lothringen.' *Jahrbuch der Deutschen Landwirtschafts-Gesellschaft* 28 (1913): 467–487.

Landwirtschaft, Ministerium fuer. *Das landwirtschaftliche Bildungswesen in Preussen: Denkschrift*. Berlin: Parey, 1929.

Lawrence, Christopher. 'Incommunicable knowledge: science, technology and the clinical art in Britain, 1850–1914.' *Journal of Contemporary History* 20 (1985): 503–520.

——. *Medicine in the Making of Modern Britain, 1700–1920*. London: Routledge, 1994.

Lawrence, Susan. 'Medical education.' In *Companion Encyclopedia of the History of Medicine*, edited by W.F. Bynum and Roy Porter, 1151–1179. London/New York: Routledge, 1993.

Layton, Edwin. *The Revolt of the Engineers: Social Responsibility and the American Engineering Profession*. Baltimore: Johns Hopkins University Press, 1971.

——. 'American ideologies of science and engineering.' *Technology and Culture* 17 (1976): 688–700.

——. 'The dimensional revolution: the new relations between theory and experiment in engineering in the age of Michelson.' In *The Michelson Era in American Science, 1870–1930*, edited by S. Goldberg and R. Stuewer, 23–41. New York: American Institute of Physics, 1988.

Lemmermann, Otto. 'Das Studium der Landwirtschaft an den landwirt. Hochschulen und landwirt. Universitaetsinstituten.' *Illustrierte Landwirtschaftliche Zeitung*, 1 November 1919, 437–439.

Lexis, Wilhelm, ed. *Die Hochschulen fuer besondere Fachgebiete im Deutschen Reich* (= vol. 4, part 2 of Lexis, *Das Unterrichtswesen im Deutschen Reich)*. Berlin: A. Asher, 1904.

Liebig, Justus. 'Wissenschaft und Landwirtschaft I: Akademische Rede vom 26.3.61.' In *Reden und Abhandlungen*, edited by Justus Liebig, 189–196. Leipzig/Heidelberg: Wintersche Verlagshandlung, 1874.

——. 'Wissenschaft und Landwirtschaft II: Akademische Rede vom 28.11.61.' In *Reden und Abhandlungen*, edited by Justus Liebig, 202–219. Leipzig/Heidelberg: Wintersche Verlagshandlung, 1874.

Liebscher, G. 'The principles and methods of breeding cultivated plants.' *Experiment Station Record* 7 (1897): 347–360.

Lindenfeld, David. 'The professionalisation of applied economics: German counterparts to business administration.' In *The German Professions, 1800–1950*, edited by Geoffrey Cocks and Konrad Jarausch, 213–231. Oxford/New York: Oxford University Press, 1990.

——. *The Practical Imagination: the German Sciences of State in the 19th Century*. Chicago: University of Chicago Press, 1997.

Lindstrom, E.W. 'Notes on Geneticists of Europe and their Laboratories – 1927–1928' (1928). In W.E. Tisdale, E.W. Lindstrom and W.J. Robbins' logbook, RG 12.1, box 37, RF Archives, RAC.

Locke, Robert. *The End of Practical Man: Entrepreneurship and Higher Education in Germany, France and Great Britain, 1880–1940*. Greenwich, CT: JAI Press, 1984.

Loewy, Ilana. *Medical Acts and Medical Facts: the Polish Tradition of Practice-Grounded Reflections on Medicine and Science [...]* Cracow: Polish Academy of Sciences, 2000.

Lucas, Werner. *Die Berufsausbildung des landwirtschaftlichen Akademikers und seine Betaetigungsmoeglichkeiten.* Berlin: Parey, 1931.

Ludmerer, Kenneth. *Learning to Heal: the Development of American Medical Education.* New York: Basic Books, 1985.

Lundgreen, Peter. 'Engineering education in Europe and the USA, 1750–1930: the rise to dominance of school culture and the engineering professions.' *Annals of Science* 47 (1990): 33–76.

Maat, Harro. *Science Cultivating Practice: a History of Agricultural Science in the Netherlands and its Colonies, 1863–1986*. Wageningen: Ponsen and Looijen, 2001.

Mahadevan, Dushyanthan. 'Creating Tomorrow's Doctors: the Development of Problem-Based Learning in British Medical Schools.' Undergraduate dissertation, Centre for the History of Science, Technology & Medicine, University of Manchester, 2002.

Manegold, Karl-Heinz. *Universitaet, Technische Hochschule und Industrie.* Berlin: Duncker und Humblot, 1970.

——. 'Die Entwicklung der T.H. Hannover zur wissenschaftlichen Hochschule: ein Beitrag zum Thema 'Verwissenschaftlichung der Technik im 19. Jahrhundert'.' In *Geschichte der Naturwissenschaften und der Technik im 19. Jahrhundert: 6. Gespraech der Georg-Agricola-Gesellschaft*, 13–46. Duesseldorf: VDI-Verlag, 1970.

——. 'Technology academised: education and training of the engineer in the 19th century.' *Sociology of the Sciences Yearbook* 2 (1978): 137–158.

——. 'Geschichte der Technischen Hochschulen.' In *Technik und Bildung*, edited by L. Boehm and Charlotte Schoenbeck, 204–234. Duesseldorf: VDI-Verlag, 1989.

Marcus, Alan. 'The wisdom of the body politic: the changing nature of publically sponsored American agricultural research since the 1830s.' *Agricultural History* 62 (1988): 4–26.

Marsden, Ben. 'Engineering science in Glasgow: economy, efficiency and measurement as prime movers in the differentiation of an academic discipline.' *British Journal for History of Science* 25 (1992): 319–346.

Martiny, B. 'Zur Frage des hoeheren landwirtschaftlichen Unterrichts.' *Deutsche Landwirtschaftliche Presse*, 5 November 1919, 678.

——. 'Zur Frage der Beseitigung der landwirt. Universitaetsinstitute.' *Deutsche Landwirtschaftliche Presse*, 8 October 1919, 612–613.

Mauel, Kurt. 'Der VDI und seine Einflussnahme auf der Entwicklung der Ingenieursausbildung an Technischen Hochschulen und Ingenieur-Schulen.' *Humanismus und Technik* 18 (1974): 91–108.

McCook, Stuart. 'Promoting the "practical": science and agricultural modernization in Puerto Rico and Colombia, 1920–1940.' *Agricultural History* 75 (2001): 52–82.

McMahon, A.M. *The Making of a Profession: a Century of Electrical Engineering in America*. New York: IEEE Press, 1984.

Mendelsohn, Andrew. 'Cultures of Bacteriology: Formation and Transformation of a Science in France and Germany, 1870–1914.' PhD, Princeton University, 1996.

Meyer, K. 'Prof. Dr. F. Berkner zum 80. Geburtstag.' *Zeitschrift fuer Acker- und Pflanzenbau* 97 (1954): 261–266.

Ministry of Agriculture & Fisheries and Dept.of Agriculture for Scotland. 'Report of the Committee on Veterinary Education in Great Britain.' London: HMSO, 1938.

Mitscherlich, E.A. 'Das Studium der Landwirtschaft und die landwirtschaftliche Forschung.' *Illustrierte Landwirtschaftliche Zeitung*, 4 October 1919, 401–402.

——. *Lebenserinnerungen* (vol. 3 of *Selbstbiographien von Naturforschern)*. Halle: Deutsche Akademie der Naturforscher Leopoldina, 1945.

Moeller, Robert G. *German Peasants and Agrarian Politics, 1914–1924: the Rheinland and Westphalia*. Chapel Hill, NC & London: University of North Carolina Press, 1986.

Mommsen, Christian. *Geheimrat Hansen. Sein Leben und sein Werk*. Frankfurt a.M.: DLG-Verlag, 1965.

Moraw, Peter. 'Organisation und Lehrkoerper der Ludwigs-Universitaet Giessen in der ersten Haelfte des 20. Jahrhunderts.' In *Giessener Gelehrte in der ersten Haelfte des 20. Jahrhunderts*, edited by Hans Georg Gundel, Peter Moraw and Volker Press, 24–75. Marburg: N.G. Elwert, 1982.

——. 'Humboldt in Giessen.' *Geschichte und Gesellschaft* 10 (1984): 47–71.

Morrell, Jack. *Science at Oxford, 1914–1939: Transforming an Arts University*. Oxford: Oxford University Press, 1997.

Mueller, H-H, and Volker Klemm. *Im Dienste der Ceres: Streiflichter zu Leben und Werk bedeutenden deutscher Landwirte und Wissenschaftler*. Leipzig/Jena/Berlin: Urania, 1988.

Muensterberg, Hugo. *The Americans*. London: Williams and Norgate, 1905.

Muenzinger, Adolf. *Aus meinem Leben*. Stuttgart: Eugen Ulmer, 1963.
Nachmansohn, David. *German-Jewish Pioneers in Science, 1900–1933*. Berlin/Heidelberg/New York: Springer-Verlag, 1979.
Natho, G. and E.-M. Wiedenroth. 'Zur Geschichte der Botanik an der Landwirtschaftlich-Gaertnerischen Fakultaet der Humboldt-Universitaet zu Berlin.' *Wissenschaftliche Zeitschrift der Humboldt-Universitaet zu Berlin, Mathe.-Naturw. Reihe* 34 (1985): 235–245.
Nathusius, Lilly. *Theodor Roemer: Lebensabriss und bibliographischer Ueberblick*. Halle: Universitaets- und Landesbibliothek, 1955.
Neuberg, Carl. 'Biochemie.' In *Wissenschaft und Landwirtschaft. Festschrift zum 50-jaehrigen Bestehen der Landwirtschaftlichen Hochschule Berlin*, edited by Friedrich Aereboe, 155–160. Berlin: Parey, 1931.
Offer, Avner. *The First World War: an Agrarian Interpretation*. Oxford: Clarendon, 1989.
Olby, Robert. 'The Development Commission and state support for applied biology in Edwardian Britain.' *Annals of Science* 48 (1991): 509–526.
Oldenburg, Gustav. *Das landwirtschaftliche Unterrichtswesen im Koenigreich Preussen[...]*. Berlin, 1913.
——. *Der Ausbau des landwirtschaftlichen Unterrichts- und Beratungswesens in Preussen*. Berlin: Parey, 1920.
Opitz, Kurt. 'Das Institut fuer Acker- und Pflanzenbau an der landwirt. Hochschule Berlin.' *Deutsche Landwirtschaftliche Presse* 1926, no. 42, 527.
——. 'Acker- und Pflanzenbau.' In *Wissenschaft und Landwirtschaft. Festschrift zum 50-jaehrigen Bestehen der Landwirtschaftlichen Hochschule Berlin*, edited by Friedrich Aereboe, 161–168. Berlin: Parey, 1931.
Overfield, Richard. *Science with Practice: Charles E. Bessey and the Maturing of American Botany*. Ames, IO: Iowa State University Press, 1993.
Palladino, Paolo. 'Between craft and science: plant breeding, Mendelian genetics, and British universities, 1900–1920.' *Technology and Culture* 34 (1993): 300–323.
——. *Entomology, Ecology and Agriculture: the Making of Scientific Careers in North America, 1885–1985*. Amsterdam: Harwood Academic, 1996.
Papendieck, Christine. 'Die Entwicklung der akademischen landwirtschaftlichen Ausbildung in Berlin von der Gruendung der Landwirt. Hochschule bis zur Bildung der agrarwissenschaftlichen Sektionen an der Humboldt Universitaet.' PhD, Humboldt-Universitaet zu Berlin, 1983.

Perkins, John A.. 'The organisation of German industry, 1850–1930: the case of beet-sugar production.' *Journal of European Economic History* 19 (1990): 549–574.

Perkins, John H.. *Geopolitics and the Green Revolution: Wheat, Genes and the Cold War*. New York: Oxford University Press, 1997.

Pollard, A.F. *The Education and Training of Chartered Engineers for the 21st Century*. London: Fellowship of Engineering, 1989.

Radkau, Joachim. *Technik in Deutschland: vom 18. Jahrhundert bis zur Gegenwart*. Frankfurt a.M.: Suhrkamp, 1989.

Raum, Hans. 'Die landwirtschaftliche Hochschullehre in Bayern, 1804–1954.' In *Bayerische Agrargeschichte*, edited by Alois Schloegl, 473–496. Munich: Bayerischer Landwirtschafts-Verlag, 1954.

——. *Beitraege zur Geschichte von Weihenstephan*, no. 4 [untitled collection of essays]. 1957.

——. 'Nachtraege zu Carl Fraas und zur Geschichte der landwirtschaftlichen Hochschullehre.' *Bayerisches landwirtschaftliches Jahrbuch* 34 (1957): 121–126.

——. 'Die Ereignisse seit 1919.' In *Beitraege zur Geschichte von Weihenstephan*, no. 5, 1958.

——. 'Die akademischen Lehrgestalten der bayerischen Landwirtschaft.' *Bayerisches landwirtschaftliches Jahrbuch* 36 (1959): 741–761.

——. 'Die landwirtschaftliche Abteilung der T.H. Muenchen von 1872–1901.' *Bayerisches landwirtschaftliches Jahrbuch* 39 (1962):745–765.

——. 'Weihenstephan unter Lintner und Kraus, 1880–1901.' *Bayerisches landwirtschaftliches Jahrbuch* 39 (1962): 842–864.

——. 'Die Doktorpromotionen, Ehrendoktorate und Habilitationen fuer Landwirte in Bayern, 1905–1950.' In *Beitraege zur Geschichte von Weihenstephan*, no. 9, 1966.

——. 'Lebenserinnerungen.' In *Beitraege zur Geschichte von Weihenstephan*, no. 10, 1967.

——. 'Die von 1951–1972 verstorbenen Professoren.' In *Beitraege zur Geschichte von Weihenstephan*, no. 13, 1973.

Redlich, Fritz. 'Academic education for business: its development and the contribution of Ignaz Jastrow.' In *Steeped in Two Cultures*, edited by Fritz Redlich, 199–257. New York: Harper and Row, 1971.

Reich, Leonard. *The Making of American Industrial Research: Science and Business at General Electric and Bell, 1876–1926*. Cambridge: Cambridge University Press, 1985.

Reichrath, Susanne. *Entstehung, Entwicklung und Stand der Agrarwissenschaften in Deutschland und Frankreich.* Frankfurt a.M.: Peter Lang, 1991.

Reiner, Hans. 'Das bayerische Hochschulwesen in den Jahren 1913/14 und 1921/22 unter Beruecksichtigung seiner entwicklung seit 1826/27.' *Zeitschrift des Bayerischen Statistischen Landesamts* 55 (1923): 98–113.

Reingold, Nathan and J. Molella. 'Introduction: the interaction of science and technology in the industrial age.' *Technology and Culture* 17 (1976): 624–633.

Reinhold, Gerhard and Max Rolfes. 'Zur Geschichte der Landwirtschaftswissenschaft an der Ludwigs-Universitaet und der Justus-Liebig-Hochschule in Giessen'.' In *Ludwigs-Universitaet-Justus-Liebig-Hochschule, 1607–1957*, 347–367. Giessen: von Muenchowsche Universitaets-druckerei Wilhelm Schmitz, 1957.

Remy, Theodor. 'Das hoehere landwirtschaftliche Unterrichtswesen und die landwirtschaftliche Forschung.' *Illustrierte Landwirtschaftliche Zeitung*, 22 November 1919, 461–462.

——. 'Landwirtschaftliche Hochschulen oder landwirtschaftliche Universitaetsinstitute?' *Deutsche Landwirtschaftliche Presse*, 1 November 1919, 669–670.

——. 'Die landwirtschaftliche Akademie in Bonn-Poppelsdorf als selbststaendige landwirtschaftliche Hochschule oder als Universitaetsfakultaet.' *Koelnische Volkszeitung und Handelsblatt*, 10 September 1919 1919.

Renner, Kurt. *Zur Geschichte des landwirtschaftlichen Bildungswesens.* Pfaffenweiler: Centaurus, 1992.

Renner, Otto. [review of Fruwirth, *Handbuch der landwirt. Pflanzenzuechtung*, vol 1]. *Biologisches Zentralblatt* 34 (1914): 555.

Richards, Stewart. '"Masters of arts and bachelors of barley": the struggle for agricultural education in mid-19th century Britain.' *History of Education* 12 (1983): 161–175.

——. 'The South-Eastern Agricultural College and public support for technical education, 1894–1914.' *Agricultural History Review* 36 (1988): 172–187.

——. *Science, Agriculture and the Universities: Changing Relations through Two Hundred Years*. unpublished book typescript, n.d. (late 1980s).

Ries, Ludwig-Wilhelm. 'Zum hundertsten Geburtstag Friedrich Aereboes.' *Zeitschrift fuer Agrargeschichte und Agrarsoziologie* 13 (1965): 44–61.

Ringer, Fritz K. *Decline of the German Mandarins: the German Academic Community, 1890–1933*. Cambridge, MA: Harvard University Press, 1969.

——. 'A sociography of German academics, 1863–1938.' *Central European History* 25 (1993): 251–280.

Rintelen, Paul. 'Th. Brinkmann (1877–1951).' In *Grosse Landwirte*, edited by Guenther Franz and Heinz Haushofer, 369–377. Frankfurt a.M.: DLG Verlag, 1970.

Roemer, Theodor. 'Variabilitaetsstudien – Dissertation.' *Archiv fuer Rassen- und Gesellschaftsbiologie* 7 (1910): 397–469.

——. 'Begleitwort.' *Zeitschrift fuer Pflanzenzuechtung* 3 (1915): 265–266.

——. *Beobachtungen auf dem Gebiete des Ackerbaues in den Vereinigten Staaten von Nordamerika*. Berlin: Parey, 1926.

——. 'Entwicklungslinien des deutschen Getreidebaus: die deutsche Landwirtschaft unter volks- und weltwirtschaftlichen Gesichtspunkte.' *Berichte ueber Landwirtschaft* suppl. vol. 50 (1932).

Rolfes, Max. 'Friedrich Aereboe (1865–1942).' In *Grosse Landwirte*, edited by Guenther Franz and Heinz Haushofer, 321–332, 1970.

Rosenberg, Charles. *No Other Gods: on Science and American Social Thought*. Baltimore/London: Johns Hopkins University Press, 1976.

——. 'Toward an ecology of knowledge: on discipline, context and history.' In *The Organization of Knowledge in Modern American, 1860–1920*, edited by A. Oleson and John Voss, 440–455. Baltimore/London: Johns Hopkins University Press, 1979.

Rossiter, Margaret. *The Emergence of Agricultural Science: Justus Liebig and the Americans, 1840–1880*. New Haven: Yale University Press, 1975.

——. 'The organization of the agricultural sciences.' In *The Organization of Knowledge in Modern America, 1860–1920*, edited by A. Oleson and John Voss, 211–248. Baltimore: Johns Hopkins University Press, 1979.

——. 'A List of References for the History of Agricultural Science in America.' Typescript, University of California, Davis, 1980.

Rothstein, William. *American Medical Schools and the Practice of Medicine: a History*. New York/Oxford: Oxford University Press, 1987.

Royal College of Veterinary Surgeons. *Report of the Committee of Inquiry into Veterinary Research*. London: The Wellcome Trust, 1997.

Ruemker, Kurt von. *Die Ausbildung des praktischen Landwirts: ein Mahnwort an alle Landwirte und solche die es werden wollen*. Berlin: Parey, 1896.

——. 'Gehoert die Landwirtschaftslehre an die Universitaeten?' *Zeitschrift der Landwirtschaftskammer fuer die Provinz Schlesien* 1, no. 8 (1897): 181–184 and 213–217.
——. 'Die moderne Landwirtschaftswissenschaft und ihre Vertretung an den Universitaeten.' *Journal fuer Landwirtschaft* 45 (1897): 335–392.
——. 'Der wirtschaftliche Mehrwert guter Kulturvarietaeten und auserlesenen Saatgutes.' *Arbeiten der Deutschen Landwirtschafts-Gesellschaft* 36 (1898): 127–153.
——. 'Ueber die Bedeutung der landwirt. Versuchsstationen und speciell ueber den Wert und die Organisation von Versuchswirtschaften in historischer Beleuchtung.' *Journal fuer Landwirtschaft* 48 (1900): 179–203.
——. 'Programm fuer das Studium der Landwirtschaft an der Koeniglichen Universitaet Breslau.' *Mitteilungen der landwirtschaftlichen Institute der Koeniglichen Universitaet Breslau* 2, no. 1 (1902): 1–23.
——. 'Erster Bericht ueber das landwirtschaftliche Versuchsfeld der Koeniglichen Universitaet Breslau in Rosenthal.' *Mitteilungen der landwirtschaftlichen Institute der Koeniglichen Universitaet Breslau* 2 (1904): 817–911.
——. *Landwirtschaft und Wissenschaft: ein offenes Wort zur Klaerung der Lage*. Berlin: Parey, 1905.
——. 'Korrelative Veraenderungen bei der Zuechtung des Roggens nach Kornfarbe.' *Jahrbuch fuer landwirtschaftliche Pflanzen- und Tierzucht* 2 (1905): 69–78.
——. 'Was koennen wir aus der Geschichte fuer die Zukunft der Landwirtschaftswissenschaft lernen?' *Fuehlings Landwirtschaftliche Zeitung* 54 (1905): 709–730.
——. 'Zur Frage der Ausgestaltung des hoechsten landwirtschaftlichen Unterrichtswesen.' *Hochschulnachrichten* 16 (1906): 89–91.
——. *Methoden der Pflanzenzucht in experimenteller Pruefung*. Berlin: Parey, 1909.
——. 'Organisation und Ziele des landwirtschaftlichen Studiums in Breslau.' *Mitteilungen der landwirtschaftlichen Institute der Koeniglichen Universitaet Breslau* 4, no. 1 (1909): 9–34.
——. 'Ueber die Entwicklung der Landwirtschaftlichen Wissenschaft in der Vergangenheit nebst einem Blick in die Zukunft.' *Deutsche Landwirtschaftliche Presse*, 18 Dec and 22 Dec 1909, 1073–1075 and 1091–1092.
——. 'Landeskulturhochschulen?' *Deutsche Landwirtschaftliche Presse*, 24 Nov. 1909.

——. *Ueber Organisation der Pflanzenzuechtung*. Berlin: Parey, 1909.
——. 'Breslau.' In *Die deutsche landwirtschaftliche Pflanzenzucht*, edited by Paul Hillmann, 24–46. Berlin: Deutsche Landwirtschafts-Gesellschaft, 1910.
——. 'Was koennen wir von der Saatzucht des Auslandes lernen?' *Jahrbuch der Deutschen Landwirtschafts-Gesellschaft* 25 (1910): 47–59.
——. *Das landwirtschaftliche Versuchs- und Unterrichtswesen in den USA*. Berlin: Parey, 1911.
——. 'Die Systematik und Methodik der modernen landwirtschaftlichen Pflanzenproduktionslehre.' *Fuehlings Landwirtschaftliche Zeitung* 60 (1911): 409–421.
——. 'Ueber Bedeutung und Methoden der Saatgutzucht.' *Mentzel und von Lengerkes Schreibkalender* (1911): 64–90.
——. 'Die Abteilung fuer Pflanzenzucht am Institut fuer landwirt. Pflanzenproduktionslehre der Koeniglichen Universitaet Breslau.' *Deutsche Landwirtschaftliche Presse* 1912, 75–78.
——. 'Die Entwicklung der landwirtschaftlichen Pflanzenzucht in Deutschland und ihre betriebswirtschaftliche Aufgaben.' *Zeitschrift fuer Pflanzenzuechtung* 1 (1912–13): 329–346.
——. *Die Entwicklung der Landwirtschaft in den letzten 25 Jahren*. Berlin: Parey, 1913.
——. 'Das landwirtschaftliche Versuchs- und Bildungswesen.' *Journal fuer Landwirtschaft* 62 (1914): 97–114.
——. 'Derzeitige Stand des landwirtschaftlichen Unterrichtswesens in Deutschland.' *Internationale Agrartechnische Rundschau* (1915): 615–635.
——. *Mein Leben*. 3 vols. Berlin: typescript, 1937.
Ruemker, Kurt von and Erich von Tschermak. *Landwirtschaftliche Studien in Nordamerika mit besonderer Beruecksichtigung der Pflanzenzuechtung*. Berlin: Parey, 1910.
Ruske, Walter. *Reichs- und Preussische Landesanstalten in Berlin*. Berlin: Bundesanstalt fuer Material-Pruefung, 1973.
Russell, Edmund. *War and Nature: Fighting Humans and Insects with Chemicals from World War I to Silent Spring*. Cambridge: Cambridge University Press, 2001.
Russell, Nicholas. *'Like Engend'ring Like': Heredity and Animal Breeding in Early Modern England*. Cambridge: Cambridge University Press, 1986.
Scheibe, Arnold. 'Die Bedeutung der wissenschaftlichen Institute fuer die private Pflanzenzucht.' *Bericht ueber Landwirtschaft* suppl. vol. 200 (1987).

Schiemann, Elisabeth. 'Erwin Baur.' *Berichte der Deutschen Botanischen Gesellschaft* 52 (1935): 51–114.

Schiera, Pierangelo. *Laboratorium der buergerlichen Welt. Deutsche Wissenschaft im 19. Jahrhundert*. Frankfurt a.M.: Suhrkamp, 1992.

Schindler, Franz. 'Ueber Ziele, Aufgaben und Methodik der Pflanzenproduktionslehre.' *Journal fuer Landwirtschaft* 46 (1898): 237–254.

——. *Der Getreidebau auf wissenschaftlicher und praktischer Grundlage*. Berlin: Parey, 1909.

Schling-Brodersen, Ursula. *Entwicklung und Institutionalisierung der Agrikulturchemie im 19. Jahrhundert: Liebig und die Landwirtschaftliche Versuchsstationen*. Braunschweig: Technische Universitaet Braunschweig, 1989.

Schloegl, Alois. *Bayerische Agrargeschichte: die Entwicklung der Land- und Forstwirtschaft seit Beginn des 19. Jahrhunderts*. Munich: Bayerischer Landwirtschafts-Verlag, 1954.

Schnabel, Franz. *Deutsche Geschichte im 19. Jahrhundert, Band 3 – Erfahrungswissenschaften und Technik*. 3rd ed. Freiburg i.Br., 1954.

Scholz, Hartmut. 'Friedrich Althoffs Einfluss auf die Entwicklung der Chemie in Deutschland.' In *Wissenschaftsgeschichte und Wissenschaftspolitik im Industriezeitalter. Das 'System Althoff' in historischer Perspektive*, edited by Bernhard vom Brocke, 337–354. Hildesheim: Verlag August Lax, 1991.

Schriefers, Herbert. 'Eduard Buchner.' In *Dictionary of Scientific Biography*, edited by C.C. Gillispie, 560–563. New York: Scribners, 1971.

Schroeder, [?]. 'Hochschule oder Universitaetsinstitut?' *Illustrierte Landwirtschaftliche Zeitung*, 25 October 1919, 429.

Schroeder, Otto. *Das Studium der Landwirtschaft und verwandter Betriebe auf den Universitaeten und Hochschulen Deutschlands[...]*. Halle: Buchhandlung des Waisenhauses, 1927.

Schwabe, Heinz. 'Zur Idee des *universitas litterarum* bei Julius Kuehn und seinen Nachfolgern.' In *Hundert Jahre landwirtschaftliche Institute der Universitaet Halle*. Halle: Landwirtschaftliche Fakultaet der Martin-Luther-Universitaet, 1963.

Seelhorst, C. [review of Fruwirth, *Wie kann sich der Landwirt Pflanzenzucht[...]zu Nutze machen?*]. *Journal fuer Landwirtschaft* 56 (1908): 114.

——. [review of Hansen, *Das landwirt. Unterrichtswesen* (1st ed.)]. *Deutsche Landwirtschaftliche Presse* 46 (1919): 299–300.

——. 'Das landwirtschaftliche Unterrichtswesen und die Ausbildung des Landwirts.' *Deutsche Landwirtschaftliche Presse*, 21 May 1919, 299–300.

Seely, Bruce. 'The scientific mystique in engineering: highway research at the Bureau of Public Roads, 1918–1940.' *Technology and Culture* 25 (1984): 798–831.

——. 'Research, engineering and science in American engineering colleges, 1900–1960.' *Technology and Culture* 34 (1993): 344–386.

——. 'SHOT, the history of technology and engineering education.' *Technology and Culture* 36 (1995): 739–772.

——. 'The "imbalance" of theory and practice in American engineering education: reforms and changes, 1920–1980.' *Icon* 5 (1999): 40–63.

Servos, John. 'The industrial relations of science: chemical engineering at MIT, 1900–1939.' *Isis* 71 (1980): 531–549.

——. 'Engineers, businessmen and the academy: the beginnings of sponsored research at the University of Michigan.' *Technology and Culture* 37 (1996): 721–762.

Settegast, Hermann. *Erlebtes und Erstrebtes*. Berlin: Puttkammer und Muehlbrecht, 1892.

Shepardson, Whitney H. *Agricultural Education in the United States*. New York: Macmillan, 1929.

Shepherd, Herbert. 'The engineer and his culture.' *Explorations in Entrepreneurial History* 4 (1952): 211–218.

Shinn, Terry. 'From "corps" to "profession": the emergence and definition of industrial engineering in modern France.' In *The Organisation of Science and Technology in France, 1808–1914*, edited by Robert Fox and George Weisz, 183–208. Paris/Cambridge: Maison des Sciences de l'Homme and Cambridge University Press, 1980.

——. 'Reactionary technologists: the struggle over the Ecole Polytechnique, 1880–1914.' *Minerva* 22 (1984): 329–345.

Sinclair, Bruce. 'Harvard, MIT and the ideal technical education.' In *Science at Harvard: Historical Perspectives*, edited by Clark Elliott and Margaret Rossiter, 76–95. Bethlehem, PA: Lehigh University Press, 1992.

Sivin, Nathan. 'Why the Scientific Revolution did not take place in China – or didn't it?' In *Transformation and Tradition in the Sciences*, edited by Everett Mendelsohn, 531–554. Cambridge: Cambridge University Press, 1984.

Skibbe, Bruno. 'Zur Vorgeschichte der Landwirtschaftlich-Gaertnerischen Fakultaet (1806–1881).' *Wissenschaftliche Zeitschrift der Humboldt-*

Universitaet zu Berlin. Beiheft z. Jubilaeums-Jahrgang 9 (1959–60): 229–256.
Sohn, Friedrich. 'Das landwirtschaftliche Beratungswesen in den Vereinigten Staaten.' *Illustrierte Landwirtschaftliche Zeitung* 11 October 1929 (1929): 471–472.
Soxhlet, Franz. 'Denkschrift betreffend die landwirt. Abteilung der Koeniglich T.H. Muenchen.' Munich (privately printed), 1896.
Spaeth, Manfred. 'Der Ingenieur als Buerger: Frankreich, Deutschland und Russland im Vergleich.' In *Buergerliche Berufe: zur Sozialgeschichte der freien und akademischen Berufe im internationalen Vergleich*, edited by Hannes Siegrist, 84–105. Goettingen: Vandenhoeck und Ruprecht, 1988.
Spiewak, Martin. 'Klassenkampf: die Fachhochschulen werden weiterhin diskriminiert.' *Die Zeit*, 23 May 2002.
Springer, E. 'Die Koeniglich Wuerttembergische landwirtschaftliche Akademie Hohenheim.' In *Das Unterrichtswesen im Deutschen Reich*, edited by Wilhelm Lexis, 96–102. Berlin: A. Asher, 1904.
Staudenmaier, John. *Technology's Storytellers: Reweaving the Human Fabric.* Cambridge, MA: MIT Press, 1985.
Steiner, Adolf Martin. 'Zur Geschichte der Saatgutpruefung und Saatgutkunde in Hohenheim.' *Der Goldene Pflug (Deutsches Landwirt. Museum, Universitaet Hohenheim)* 5 (1995): 12–15.
Steuert, Ludwig. *Die Koenigliche Bayerische Akademie Weihenstephan und ihre Vorgeschichte. Festschrift zur Jahrhundertfeier 1905.* Berlin: Parey, 1905.
Stokes, Donald. *Pasteur's Quadrant: Basic Science and Technological Innovation.* Washington, DC: Brookings Institution Press, 1997.
Storer, Norman. *Science and Scientists in an Agricultural Research Organization; a Sociological Study.* New York: Arno Press, 1980.
Sturdy, Steve. 'The political economy of scientific medicine: science, education and the transformation of medical practice in Sheffield, 1890–1922.' *Medical History* 36 (1992): 125–159.
Stutzer, A. and Paul Gisevius. *Der Wettbewerb der daenischen und der schwedischen Landwirte mit Deutschland. Reiseerinnerungen.* Stuttgart: Eugen Ulmer, 1904.
Tamm, Ernst. 'Prof. Dr. Kurt Opitz: 70 Jahre.' *Der Zuechter* 17/18 (1947): 449.
Teich, Mikulas. *Bier, Wissenschaft und Wirtschaft in Deutschland 1800–1914.* Vienna: Boehlau, 2000.

Titze, Hartmut, ed. *Datenhandbuch zur deutschen Bildungsgeschichte*. Vol. I, part 1. Goettingen: Vandenhoeck und Ruprecht, 1987.
Tribe, Keith. *Strategies of Economic Order*. Cambridge: Cambridge University Press, 1994.
——. 'Business education at the Mannheim Handelshochschule, 1907–1933.' *Minerva* 32 (1994): 158–185.
Trischler, Helmuth. 'Self-mobilization or resistance? aeronautical research and National Socialism.' In *Science, Technology and National Socialism*, edited by Monika Renneberg and Mark Walker, 72–87. Cambridge: Cambridge University Press, 1994.
Tschermak, Erich von. 'Zum 60. Geburtsfeste Prof. Dr. Carl Fruwirths.' *Zeitschrift fuer Pflanzenzuechtung* 8 (1922): 324–330.
——. 'Ein Leben fuer die Zuechtung: aus der Werkstatt eines alten Pflanzenzuechters.' *Odal: Monatshefte fuer Blut und Boden* 10 (1941): 768–777.
——. *Leben und Wirken eines oesterreichischen Pflanzenzuechters*. Berlin/Hamburg: Parey, 1958.
Ullmann, Hans-Peter. *Interessenverbaende in Deutschland*. Frankfurt a.M.: Suhrkamp, 1988.
Ullrich, Hermann. 'Friedrich August Koernicke (1828–1908).' In *Bonner Gelehrte: Beitraege zur Geschichte der Wissenschaft in Bonn*, 87–97. Bonn: Bouvier/Roehrscheid, 1971.
Valier, Helen. 'The Politics of Scientific Medicine in Manchester, ca. 1900–1960.' PhD, University of Manchester, 2002.
Vernon, Keith. 'Pus, sewage, beer and milk: microbiology in Britain, 1870–1940.' *History of Science* 28 (1990): 289–325.
——. 'Science for the farmer? agricultural research in England, 1909–1936.' *Twentieth Century British History* 8 (1997): 310–333.
Vessuri, Hebe. 'Foreign scientists, the Rockefeller Foundation and the origins of agricultural science in Venezuela.' *Minerva* 32 (1994): 267–296.
Vossler, O. von. *Die Entwicklung Hohenheims in den letzten Jahrzehnten*. Plieningen: Find, 1893.
Wacker, Johann. 'Die Kgl. wuerttembergische Saatzuchtanstalt Hohenheim.' In *Die deutsche landwirtschaftliche Pflanzenzucht*, edited by Paul Hillmann, 116–135. Berlin: Deutscher Landwirtschafts-Gesellschaft, 1910.
——. 'Die landwirtschaftliche Pflanzenzucht und ihre land- und volkswirtschaftliche Bedeutung.' *Zeitschrift fuer Pflanzenzuechtung* 9 (1923–24): 35–49.

Wagner, Fritz. 'Zur Frage des Studiums der praktischen Landwirte.' *Deutsche Landwirtschaftliche Presse* 47 (1920): 92.

Warner, John Harley. *The Therapeutic Perspective: Medical Practice, Knowledge, and Identity in America, 1820–1885*. Cambridge, MA/London: Harvard University Press, 1986.

——. 'Ideals of science and their discontents in late 19th century American medicine.' *Isis* 82 (1991): 454–478.

——. 'The history of science and the sciences of medicine.' *Osiris* 10 (1995): 164–193.

Waterstradt, Franz. 'Aufgaben der Landwirtschaftswissenschaft und ihrer Lehre.' *Fuehlings Landwirtschaftliche Zeitung* 63 (1914): 318–327.

Weiss, John. *The Making of Technological Man: the Social Origins of French Engineering Education*. Cambridge, MA: MIT Press, 1982.

Welte, Erwin. 'Die Bedeutung der mineralischen Duengung und die Duengemittelindustrie in den letzten 100 Jahren.' *Technikgeschichte* 35 (1968): 37-55.

White, Fred C. 'The interrelationships of agricultural research and farm structure.' In *The Agricultural Scientific Enterprise: a System in Transition*, edited by Lawrence Busch and William Lacey, 255–264. Boulder, CO/London: Westview Press, 1986.

Wiedemann, Fr. 'Wissenschaft und Praxis.' *Deutsche Landwirtschaftliche Presse*, 23 October 1926, 545–546.

Wieland, Thomas. '"Wir beherrschen den pflanzlichen Organismus besser,...": Akademische Pflanzenzuechtung in Deutschland, 1889–1945.' PhD, Universitaet Muenchen, 2000.

Williams, Rosalind. '"All that is solid melts into air": historians of technology in the information revolution.' *Technology and Culture* 41 (2000): 641–668.

Wissenschaftsrat. 'Empfehlungen zur Entwicklung der Fachhochschulen.' Berlin, 2002.

Wittmack, Ludwig, ed. *Die Koenigliche Landwirtschaftliche Hochschule in Berlin. Festschrift zur Feier des 25-jaehrigen Bestehens*. Berlin: Parey, 1906.

Woelfer, [?]. 'Nochmals: "Das landwirtschaftliche Unterrichtswesen und die Ausbildung des Landwirts."' *Deutsche Landwirtschaftliche Presse*, 24 May 1919, 311.

Woelfer, [?] and Lentz. "Landwirtschaftliche Hochschulen oder landwirtschaftliche Universitaetsinstitute?" *Deutsche Landwirtschaftliche Presse*, 29 October 1919, 660.

Wohltmann, Friedrich. *Landwirtschaftliche Reisestudien ueber Chicago und Nord-Amerika*. Breslau: Schletter, 1894.

——. 'Das landwirtschaftliche Institut der Universitaet zu Halle a.S., seine Entwicklung und Neugestaltung.' *Kuehn-Archiv* 1 (1911): 1–59.

——, ed. *Festschrift zum Feier des 50-jaehrigen Bestehens des Landwirtschaftlichen Instituts (= Kuehn-Archiv,* Vol. 5*)*. Berlin: Parey, 1914.

——. 'Die Festfeier des 50-jaehrigen Bestehens des Landwirt. Instituts der Universitaet Halle-Wittenberg.' *Kuehn-Archiv* 6 (1915): 1–32.

Wolf, J. 'Die Koenigliche Universitaet Breslau.' In *Das Unterrichtswesen im Deutschen Reich*, edited by Wilhelm Lexis, 347–362. Berlin: A. Asher, 1904.

Wolff. 'Ferdinand Wohltmann und die deutsche Landwirtschaft.' *Jahrbuch der Deutschen Landwirtschafts-Gesellschaft* 36 (1921): 126a–126d.

Wollny, E. and Emil Pott. 'Antrag betreffend die Verlegung der landwirt. Abteilung der Koeniglichen T.H. Muenchen an die Universitaet daselbst.' Munich (privately printed), 1896.

Wood, Roger and Vitezslav Orel. *Genetic Prehistory in Selective Breeding: a Prelude to Mendel*. Oxford/New York: Oxford University Press, 2001.

Wunderlich, Gustav. 'Carl Fruwirth (1862–1930).' In *Grosse Landwirte*, edited by Guenther Franz & Heinz Haushofer, 306–320, Frankfurt a.M.: DLG-Verlag, 1970.

Wygodzinski, Prof.Dr. 'Universitaeten und Fachhochschulen.' *Niedersaechsische Hochschulzeitung*, 29 October 1919, 6–7.

Zoerner, Hans. 'Reform des landwirtschaftlichen Studienwesens.' *Illustrierte Landwirtschaftliche Zeitung*, 15 May 1920, 184–186.

Zoerner, Hans and Fr. Aereboe. 'Landwirtschaftliches Unterrichtswesen.' In *Handwoerterbuch der Staatswissenschaften*, edited by Ludwig Elster et al., 196–206. Jena: G. Fischer, 1925.

Zorn, Wilhelm. 'Die neuen Landwirtschaftlichen Institute an der Universitaet Breslau.' *Illustrierte Landwirtschaftliche Zeitung* 21 April (1923): 128–129.

——. *Die Geschichte der Landwirtschaftswissenschaften in Schlesien*. Wuerzburg: Holzner, 1964.

Zweckbronner, Gerhard. 'Die Ingenieurwissenschaften im 19. Jahrhundert.' In *Festschrift zum 150-jaehrigen Bestehen der Universitaet Stuttgart*, edited by Johann H. Voigt, 189–222. Stuttgart: Deutsche Verlagsanstalt, 1979.

Index

www.ingramcontent.com/pod-product-compliance
Ingram Content Group UK Ltd.
Pitfield, Milton Keynes, MK11 3LW, UK
UKHW021852100926
8556IPUK00020B/1183

9 783039 102990